BEGINNING TO END THE CLIMATE CRISIS

Luisa Neubauer and Alexander Repenning

Translated by Sabine von Mering

BEGINNING TO END THE CLIMATE CRISIS

A HISTORY OF OUR FUTURE

Brandeis University Press / Waltham, Massachusetts

First edition published 2019 by J. G. Cotta'sche Buchhandlung
as *Vom Ende der Klimakrise: Eine Geschichte unserer Zukunft*
Brandeis University Press edition 2023

Manufactured in the United States of America
Designed and composed in Arno Pro and
Brandon Grotesque by Mindy Basinger Hill

LIBRARY OF CONGRESS CATALOGING-IN-PUBLICATION DATA

Names: Neubauer, Luisa-Marie, 1996– author. | Repenning, Alexander, 1989– author. | Von Mering, Sabine, translator.
Title: Beginning to end the climate crisis : a history of our future / Luisa Neubauer and Alexander Repenning ; translated by Sabine von Mering.
Other titles: Vom Ende der Klimakrise : eine Geschichte unserer Zukunft. English
Description: Brandeis University Press Edition. | Waltham, Massachusetts: Brandeis University Press, [2023] | Translation of: Vom Ende der Klimakrise : eine Geschichte unserer Zukunft, 2019. Includes bibliographical references. | Summary: "A call to action for young people to respond to the climate crisis from two of the most prominent and successful young German climate activists"— Provided by publisher.
Identifiers: LCCN 2022045507 | ISBN 9781684581474 (Paper) | ISBN 9781684581481 (e-book)
Subjects: LCSH: Environmentalism. | Global warming.
Classification: LCC GE195.N47513 2023 | DDC 333.72—dc23/eng20221219

5 4 3 2 1

COVER IMAGES *front* Luisa Neubauer, 2021, photo by Oğuz Yılmaz
back photograph of Alexander Repenning

This book was printed by Sheridan Books using Natures Book Recycled text stock with 30% post-consumer waste (PCW) content.

FOR ALL THE POSSIBILISTS OUT THERE.

AND FOR THOSE WHO WANT TO JOIN THEM.

CONTENTS

FOREWORD *Bill McKibben*

This is a very important book about the climate crisis, for several reasons.

One, it gives American audiences a real insight into how things look on the other side of the Atlantic. Both public opinion and government policy are far advanced in Europe and Germany—but that does not mean they are moving fast enough. (Indeed, in May of 2022 news came that the EU's carbon emissions had topped their prepandemic totals.) But there the opposition is mostly in the form of delay, not denial—squishier than the idiot opposition in Trumpish America, but in many ways just as effective.

Two, it gives older audiences an insight into how younger people are thinking. Greta Thunberg's 2018 school strike galvanized the world, but it especially galvanized young Europe. Luisa and Alex were already engaged in this work but they were able to ride the tsunami that Thunberg unleashed, with enormous skill that helped quickly transform climate politics. In a sense, Greta validated a kind of impatience, and put elites on notice that their old coping methods no longer sufficed—this is a very fine portrait of what happens when movements suddenly *move*.

But the third reason is that it underlines just how much is always the same, age or geography aside. Power—and the status quo is power in itself—does not budge without a shove, and that shove has to come not just from angry tweets but from committed organizing. Of all the true sentiments in this book, the truest may be: "It's a matter of organization and mobilization: any action, no matter how small, can have a big impact if it's launched at the right moment, with the right narrative, and by as many people as possible."

I have gotten to watch the way these young people organize—watched as they've helped rewrite the script for Europe. They are extraordinarily adept, and this book gives a deep and honest reflection on why that is. May they stay at it, and may we all pitch in!

PREFACE TO THE ENGLISH EDITION

Luisa Neubauer and Alexander Repenning

Hi!

We—Alex and Luisa—are so happy you are holding this book in your hands. We wrote it not only with the goal to inspire and inform, but more importantly to spread hope and inspire rage in these difficult times. If we achieved that in this book, it's all we could wish for.

It has been almost four years since the first publication of the book in German, and so much has happened since then. We are delighted that our book is now translated into English to reach a wider, global, readership.

When we started writing in 2018, we couldn't have imagined that one year later everyone would be talking about Fridays For Future. We couldn't have imagined that Luisa would end up leading the German Fridays For Future movement, travel with Greta Thunberg to meet French President Emmanuel Macron, to discuss the climate crisis with Barack Obama, and to speak at the World Economic Forum in Davos. This book ended up documenting how to start a climate movement more or less by accident.

At the beginning of 2019, Luisa and a handful of other activists started the Fridays For Future movement in Germany. Week after week we organized climate strikes across the country, growing each time until in May 2019, the climate crisis became the most important subject for German voters ahead of the European Parliament election. Meanwhile, we went to all possible stages and spaces in Germany and Europe, speaking in front of investors, industries, and politicians. In September that year, Luisa and Fridays For Future organized the largest climate demonstration in the history of the country, bringing 1.4 million people to the streets. We didn't only pressure our government to act; just a few months later Fridays For Future started an unprecedented campaign against the energy company Siemens, which resulted in the CEO of Siemens offering Luisa a seat on the board. Instead, Luisa kept organizing throughout the pandemic, all the way up the national election in 2021, when voters across the democratic spectrum considered the climate a priority.

At the same time, Alex went on to connect activists for justice, peace, and sustainability with universities across the globe as part of Right Livelihood, the organization that first brought us together. He helped bring together undergraduate students in California with anti-oil activist Nnimmo Bassey from Nigeria, and master's degree students from Mumbai in India with Boston-based nonviolent action experts like Jamila Raqib. He helped set up a lecture with lawyer and social justice activist Bryan Stevenson at a Swiss university and linked the Landless Workers Movement in Brazil with student activists in Thailand, to name just a few examples.

We wrote this book at a time when the climate crisis kept being set aside by politicians as something to deal with in the future, to address only after other, "more pressing" issues were dealt with. The climate justice movement we're a part of has since been successful in raising awareness among the general public, among decision makers, and, at first glance, even within the industrial and financial sectors. But building awareness is not enough to avert the unfolding climate crisis. We need to build a power base, and we need to call out false climate solutions.

We saw the greenwashing attempted by multinational corporations like Nestlé, with their claims of aiming to become "net zero" in the coming decades.[1] We saw oil-giants like Royal Dutch Shell pledging to offset emissions and investing hundreds of millions of dollars into so-called "nature-based solutions," instead of changing their business model and radically reducing emissions.[2] We saw billionaires getting richer and going into outer space while still claiming to care about this fragile planet and its inhabitants. Seeing this we are reminded to remain vigilant as the very same companies and people making billions off the destructive status quo now claim to be at the forefront of fixing the problem they played a large role in creating.

We were aware of our privileged position as two students from the urban white middle class, back then in our early and late twenties, living in one of the richest societies on this planet. We had our peers in mind as we explored and explained the dimensions of the climate crisis in this book: people who were not seeing the reality of the changing climate in their everyday lives, who didn't seem to be threatened or intimidated by the scale of this threat.

That said, our book is meant for everyone, for all generations, from teenagers to grandparents. Some of the language we use might be unfamiliar, for instance that we speak of a "we" a lot. By that, we mean to include anyone who feels that real change toward justice isn't just nice, or necessary, but that it is possible.

Our calls to action were directed at decision makers and citizens around us in Germany and Europe, but we're convinced that our arguments and our actions are as relevant for the public debate in the US, the UK, Australia, New Zealand, indeed everywhere in our global community.

When we became part of kick-starting the Fridays For Future youth climate movement in 2019, something like a pandemic was considered a matter of dystopian movies, and a full-scale war in Europe like the one we are witnessing now in 2022 seemed absurd. As the twenty-first century unfolds, crises everywhere increasingly escalate and interact, pushing planetary boundaries and societal resilience beyond their limits. And that is precisely what this book is all about. It is about the intersections and struggles of a world that seems to be falling apart. And what it takes to move away from the age of crises. In times when many people have heard about movements like Fridays For Future, Sunrise, or Black Lives Matter, it is easy to forget what it takes to get things moving. And it is just as easy to assume that someone else will get started, that someone else will take care of it, someone else will feel responsible. It is so inconvenient to realize that we all—the two of us and you, who are reading this right now—that we are all that someone. And out there, there is someone who is counting on us right now to get going. How to get going and what to do—that is what we talk about in this book.

TRANSLATOR'S NOTE *Sabine von Mering*

Before 2018 I had attended many meetings in Massachusetts where I and other gray-haired climate activists deplored the absence of youth in our midst. The arrival of the Sunrise Movement in the United States, as well as Greta Thunberg and the Fridays For Future that she inspired in Europe and beyond, therefore marked a very welcome change that year. It provided a much-needed spark for the climate movement all around the globe. And as a German American climate activist and German Studies scholar I am heartened to see that Germany is home to the world's largest Fridays For Future youth climate movement today.

In November 2019, I had the opportunity to chair a panel about "Fighting the Climate Crisis" at the German-American Conference at Harvard University that included Luisa Neubauer. I remember how forcefully she articulated the need for more stringent climate policy. As I read this book in the original German, I was struck by the global connections that are woven through the text. Indeed, Luisa and Alex are familiar not only with American literature on climate activism, but also with the much larger climate justice discourses from long before Fridays For Future arrived on the scene. Their book introduces many German and other international perspectives on what we must do about the climate crisis.

Having taught a course on "Human/Nature: European Perspectives on Climate Change" at Brandeis University for over a decade, I understand the need for more relevant European texts in English translation. The present volume begins to address this need.

I am honored and pleased to be able to help create a wider platform for the vision of these two inspiring and dedicated young voices and thereby help contribute to a transatlantic conversation about the climate emergency and what we must do.

BEGINNING TO END THE CLIMATE CRISIS

INTRODUCTION

What to do when you are in the middle of a major crisis and nobody acts? How do you communicate a scientifically proven catastrophe in a time that declares itself to be post-factual? A time when 280 characters on Twitter dictate the rhythm of communication? A time when attention spans are shrinking, and an endless flow of information is rushing past you in the form of ever-faster thundering breaking news? How do you tell people about a crisis that is so dramatic that it should dominate the agenda like nothing else—but instead it is relativized, dismissed, or ignored by large segments of society and politics? How do you explain to political decision makers that they must take care of a problem that does not fit into any legislative period and is bigger than any constituency? How do you mobilize for a problem that many people don't see as a problem?

You tell stories. Personal stories. And this is our story.

Stockholm, Summer 2017.

We were dipping cinnamon cookies into our coffee—"Fika"—that's what coffee hour is called in Sweden. We were sitting in the yard behind a hundred-year-old house, the sun was shining twelve hours a day, the sky was bluer than in any travel guide. Somewhere someone was mowing the lawn, and the smell of freshly cut grass drifted over. That is where, away from the hustle and bustle of the city, the "Alternative Nobel Prize" (or the Right Livelihood Award, as it is officially called) has its offices. That the Thunberg family lives in the same city seems like a sign from today's vantage point, but at that time hardly anyone outside her family and friends knew the girl named Greta.

That Swedish summer was quite something. While the two of us were doing research for the Foundation of the Alternative Nobel Prize, the world was watching spellbound as an American president became a real-life nightmare, producing daily headlines. The Rohingya crisis in Myanmar turned hundreds of thousands of people into refugees overnight. People recalled the beginning of the global economic crisis ten years earlier while celebrating rapid economic growth at the same time. Meanwhile, many were suffering from the heat: This summer developed into one of the three hottest summers since record keeping began.

Sitting barefoot at that garden table, drinking coffee, we decided to finally make a start, asking the big questions, all summer long, every single day. We wanted to dare find answers, and not to be satisfied with a, "Well, it's pretty complicated."

How can it be that we produce enough food for over ten billion people worldwide,[1] yet over 800 million people still go hungry?[2]

What will the world look like when an additional two to three billion people live here by the middle of the century? What future awaits the over eighty million refugees across the world?[3] And the many millions who will in all likelihood be sharing their fate?

What is the explanation for the right-wing turn in Western countries that has brought nationalist parties into many parliaments and has normalized racist agitation again?

How can it be that more and more people end up burnt out, lonely, depressed to the point of hospitalization when "things have never been better"[4] here in Germany and around the world?

Many of these questions are interrelated. The crisis of all crises, however—and thus the key to everything else—is the climate crisis: How is it possible that scientists are convinced that we have been moving toward the greatest catastrophe in human history for decades, but instead of slowing down and reversing course we are actually increasing the pace?

We know that waste separation, organic vegetables, and bamboo toothbrushes are not enough when it comes to finding answers to this, one of *the* existential questions of our time. What is to be done?

The Right Livelihood Foundation, based in Stockholm, was the perfect place to begin the search for action plans fit for our future. After all, this prize has been awarded for the past forty years to people and organizations from all over the world who have found practical solutions to the global problems of our time.

Among the winners are people like Frances Moore-Lappé from the US, an author and activist who campaigns against world hunger and for democracy; Hermann Scheer from Germany, a politician who promoted solar energy worldwide as early as 1988; Vandana Shiva from India, who is engaged in ecofeminism and biodiversity; along with Yacouba Sawadogo from Burkina Faso and Tony Rinaudo from Australia, who transform deserts into forests.

We were overwhelmed by the sense of hope in the stories of these activists. At the same time the extent of the looming catastrophes rendered us speechless. We were angry that politicians are paying no attention to existing solutions, and are in fact instead actively ignoring or boycotting them. We wanted to write about that.

Back then we didn't yet realize that this would become a book about the climate crisis, because it is far from the only crisis that causes us headaches when we look toward our future. The "multiple crisis" of our time, as sociologists Markus Wissen and Ulrich Brand are calling it, encompasses all areas of our lives. Think of the ecological crisis of species extinctions, soil degradation, and environmental destruction, or the consequences of the global economic crisis, which many countries are still feeling today. Impoverishment, societal polarization, and the dismantling of the social safety nets plunge us into a "crisis of social reproduction."[5] The preservation of the achievements of social welfare states, which were supposed to secure a "life in dignity" for all, is being called into question.[6]

The increasing global refugee movements have intensified these tendencies and are nevertheless a consequence of these crises. Added to this is the crisis of representative democracy and established parties, which (with the exception of the electoral success of the Greens) became apparent again in the 2019 European election. And, of course, there is racism, inequality, and the continuing crisis of gender relations, which manifests in all levels of society—in daily sexism against women and other genders, but also structurally in the job market, in politics, in the media, and in the private sphere.

We were not in agreement as to which of these fires needed to be put out most urgently.

ALEX For me the climate crisis had always been a topic for nature lovers and those who prefer to spend time in the forest rather than with other people. I didn't have anything against them, but I felt that there were more important questions to tackle in a globalized world with all its inequalities, power differentials, and exploitative relationships than how to rewet a bog, protect a rare beetle species, or what the consequences are of changing vegetation zones. When I thought of the climate I thought that the weather in Hamburg is different from the weather in Freiburg or Palma de Mallorca, and as someone from Hamburg I was happy to occasionally see the sun shine for a change.

I first learned about the climate crisis through Al Gore's documentary film *An Inconvenient Truth*. I remember images of dried-up lakes, changing landscapes, and graphs that all rise sharply toward the end. I also remember the somewhat silly animation of a frog who doesn't leave the water while it is slowly being heated toward a deadly temperature—though it would have jumped out right away had the water been that hot at the outset. The frog, Gore said, resembles us humans, who are not reacting to the deadly threat of the climate crisis, because it reveals itself only slowly, with a time lag. But as catchy as the image of the frog was, the examples of places at the other end of the globe, which were changing due to the climate, seemed far away, with no connection to my own life. It also seemed suspect to me that this man in a suit was flying around the world and riding in limousines to lectures about the climatic consequences of our lifestyle. Thus, global warming remained a problem of distant places for me, far away as well from questions of justice and the good life for all, far away from what was on my mind at the time.

LUISA For me it was different. I first learned of the greenhouse effect at age thirteen, in geography class in school. Our teacher had planned a double lesson on the topic, two times forty-five minutes. That was it. The following week we talked about volcanoes, then about the Wadden Sea, then North America. I found it irritating that such an important topic was squeezed into just one double lesson.

At the end of the term I was left with the vague feeling that "something isn't right with our planet," and the commitment to avoid using plastic bags at the supermarket every now and then. "For the environment," people would say at the time. I began to read the *taz* [*tageszeitung*, a left-wing daily newspaper] over breakfast. The more I learned about the climate crisis, the stranger I found how it was dealt with. When I tried to become a vegetarian at age fourteen, my parents forbade it. They did not understand that this decision was a consequence of my thinking about the climate crisis. To their ears, their pubescent daughter's idea sounded above all like the first step on the path toward an eating disorder. As a compromise we had organic meat once a week, the other days I was allowed to eat vegetarian food.

One year later a friend and I started experiments on small solar panels in school. Looking back, that must have been unusual, as the energy transition was still in its infancy back then, and regenerative energies was a topic for geeks. That same year we were awarded a prize in a science competition for our rudimentary discoveries. At school I wrote essays about the environmental consequences of the Elbe River deepening, learned why solar toilets failed in Namibia, and kept writing about these things. After my graduation I started an internship with an environmental magazine. The mountain of questions kept growing—questions about the climate, about the ecological limits to growth, the future of the planet and humanity, here and in the Global South. One year later, I decided to study geography.

So, when I was sitting with Alex in that garden in Stockholm two years later, I was already looking back on many years of commitment to climate protection. Only I didn't really know what difference it had made, if any. The more we dealt with the climate crisis, the clearer it became for both of us that it would have enormous consequences for humankind. No matter whether we were dealing with questions of humane living conditions, justice, the environment, or animal protection—everything converged in the climate crisis. No matter where we began to think about the ethical tasks of our time—sooner or later we always ended up with the existential danger caused by the rising concentration of CO_2 in the atmosphere.

Back then, when the idea for this book was born, it was impossible to foresee that we young people would fill the streets worldwide as the Fridays For Future movement. I never planned to become a full-time climate activist or to use school and university strikes to demand political action. On the contrary: I had not really felt at home with environmental organizations and had never organized a demonstration. My first strike took me miles out of my comfort zone. I was afraid that nobody would show up. I did not know what to say to the people who ended up standing with me in front of the Bundestag (the federal parliament building in Berlin), freezing, or how to convince them to come back again. But we dared, and that is what this crisis demands of us: We have to step out of our comfort zone. And contrary to all my expectations, that first Friday in front of the Bundestag was the beginning of something big. From that day forward, my entire life revolved around the climate crisis.

ALARMISM? HAMBURG 2050

We both grew up in Hamburg. When the wind was blowing in the right direction, we could hear the roar of the ships' horns all the way home. The abundance of water promised us quality of life. On vacation, we boasted how Hamburg has more bridges than Amsterdam and Venice combined. The water was the essence of what made our hometown special, what we raved about when we talked about it away from home. When we reminisce about Hamburg summers today, we think of the Elbe beach, the waves, and of little sailboats with white sails.

Today, much of that has been pushed aside by concerns about proximity to water in times of climate crisis. Just a few years later, water has become a symbol of the dangers that threaten us and our children. Hamburg—like many other coastal cities—will be heavily impacted by global warming and rising sea levels. More and more frequent storm surges in the North Sea, and the growing risk of flooding from inland, will increasingly affect the city in the coming decades.[7]

All indications are that global greenhouse gas emissions will continue to rise. Depending on the calculations, the annual mean temperature is expected to rise between 2.8 and 4.7 degrees by the end of the century.[8] While this may at first sound like good news in a city notorious for its rainy days, it will most likely mean a less pleasant life. Storms and heavy downpours will increase, as will periods of heat and drought. Children will burn their feet in schoolyards in the summer. Downtown, a combination of exhaust fumes and heat waves will have deadly consequences for the elderly and the sick. Ecosystems in the city, in parks, wild areas, and around the Elbe, will collapse. Water quality will deteriorate dramatically, while farmers with fields in and around Hamburg will be struggling for their livelihoods due to unending crop failures.[9]

All these are not distant future scenarios; we will experience much of it ourselves. The climatic changes will partly destroy the places of our childhood memories, and will dominate our lives into our old age—to an extent never seen before.

For a long time, people spoke of *Enkeltauglichkeit* (in defense of future generations) in German climate policy. That no longer makes any sense: We already have to speak of *Kindertauglichkeit* (in defense of the next generation)—or even more short term, in defense of our own generation.

People often say that "we are the first to feel the climate crisis and the last who can still do something about it." This statement, too, is out of date.

It is true that we the young people are the first generation whose lives will be significantly impacted by the climate crisis in the near future. But many people are already impacted now, most heavily in places like sub-Saharan Africa. Yet the climate crisis is becoming more and more visible in privileged places in the Global North too, like in Germany. Just two examples are farmers, whose fields remain dry in summer heat, and foresters, whose forests are dying or being eaten by pests.

The second part of that statement, that we are the last generation who can still make a difference, also fails to do justice to the seriousness of the situation. Because "we" doesn't only mean the young generation, whose future is at stake due to the climate crisis. By "we" we mean the shapers of today's society. In other words, those who are helping to decide today how we will live and do business in the future. In a democracy, where everyone can participate at least at the ballot box, this means that in addition to those who are in control in politics, economics, and finance, everyone else also has a say.

WHAT DOES THE SCIENCE SAY?

We have studied the scientific explanation for the climate crisis and spoken to scientists. We have learned two things in particular:

1. The climate crisis is not just a lifestyle crisis; it does not only impact the question of how life can continue for humans and animals on a significantly changed planet.[10] The climate crisis is a question of survival on planet earth in the medium and long term. First for animals, later for human life on earth as we know it.[11]

2. If we are serious about the targets decided upon in the Paris Agree-

ment, if we do not want to allow that the earth's climate is increasingly hostile to large parts of the world's population, if we think climate protection through to the end, if we understand that climate protection is protection of humans and if we are willing to act accordingly—then all decision makers are called upon to get started immediately. We are not only the last ones who can still prevent the worst damage, we are also the ones who have to tackle this task of a century. No future society will ever be able to prevent so many dangers.

So much for the status quo.

LET'S STOP MAKING THE SAME MISTAKES OVER AND OVER AGAIN

We are not writing this book just to tell you how bad things are for the planet. NASA's homepage shows that as well. We are writing this book because we are not willing to let go of the fact that thirty years—Alex's entire lifetime—have been wasted when it comes to climate policy. We are writing because we do not want to become part of the next story about further wasted decades. Because we are not speaking about an abstract world when we speak about the year 2050, but about our own lives. We are writing this book as an appeal not only to the younger generation, but to everyone. Because we need everyone. Because it is our job to demand radical climate policies—and to enact them with all nonviolent means at our disposal.

WE ARE POSSIBILISTS

Do we look at the future with optimism? Yes and no. We take our cue from Jakob von Uexküll, the founder of the Alternative Nobel Prize. Von Uexküll's credo was to be neither an optimist nor a pessimist, but a possibilist. What is that? "The possibilist," says von Uexküll, "sees the possibilities, and it depends on each of us whether they will be realized."[12]

It is with this same attitude that we are writing this book. During our summer in Stockholm we got to know many examples that show a just, peaceful, and sustainable world is possible. What drives us is not the belief that everything will be all right, but the conviction that the catastrophe is not inescapable and much good can still be done.

We know that there are solutions to the major social problems of our time. Their implementation is not easy, and perhaps not even probable—but it is possible, and as long as this possibility exists, it is worth fighting for. We must talk about it and encourage people to become part of these solutions.

Possibilism means rolling up our sleeves. While pessimists tend to quickly fall into a paralyzing and self-pitying fatalism, and while optimists make themselves comfortable in the expectation of a rosy future, we possibilists get active. As long as there is still even a tiny chance for a better tomorrow, we should do everything we can today to take advantage of it.

Being a possibilist is uncomfortable. Getting active is exhausting. Yes, there are solutions, but they depend on mobilizing a critical mass for their realization. We must not let ourselves be distracted; not by the gloomy picture that climate science paints of the future and which leaves little room for hope, but also not by the deceptive optimism of all those who subscribe to the belief that human inventive spirit, technological progress, and the supposed therapeutic powers of the market will save us. While they continued to preach that everything would be fine, global emissions rose to record heights, and the crisis kept worsening year after year.

That is what distinguishes us possibilists from optimists and also from pessimists. We know that another future is possible, but we also know that it will not be handed to us.

AN INVITATION

We are not speaking for a "generation," whatever that may be. We are also not speaking for Fridays For Future. We are speaking for ourselves, we are telling stories from our own personal perspectives. We are inspired to do so by our experiences, by what we are learning from conversations, from our studies, on the street.

We hope that some of you will recognize yourselves in what we are writing about, and we expect that just as many will also be bothered by it. We are making an offer, and are extending an invitation to everyone to become part of the story that we will be writing from now on: It's the story of the end of the climate crisis, of the attitude with which we will confront the crisis, and of the commitment it takes.

1 OUR FUTURE IS A DYSTOPIA

LUISA Berlin, in wet and cold February. Drizzle from morning to night. Low-hanging clouds dimmed the lights on the street as if they begrudged the 69th Berlinale Film Festival that was about to begin its glamor: large posters everywhere, television crews and onlookers in front of the cinemas. Every now and then, beautiful people stepped out of big cars and hurried toward the red carpet, wearing expensive clothes that were definitely not made for this weather.

I was sitting in a movie theater chair with armrests so wide that my forearms would have fit on them three times, a dozen friends at my side, our seats marked "reserved."

Barely two months earlier we had begun to spend our Fridays in front of city halls, regional parliaments, and ministries; Fridays For Future had made headlines.

In those weeks, a so-called "coal commission" was in negotiations over when Germany should phase out coal-fired power generation. And because the commission, seemingly unimpressed by the weekly climate strikes, was about to ignore all climate targets, we called for a large strike in front of the Federal Ministry of Economics on the day of its last meeting. I had organized a permit for five hundred people, secretly hoping that we would be more than a thousand. Ten thousand came.

On the morning of the strike, the coal commission had made the seemingly conciliatory move of inviting some of us to their meeting. We were given four minutes to present our demands. We spoke of our fear of growing old on a wrecked planet, and of how—if Germany were to decide to continue to burn coal for much longer—other countries would follow our bad example. We reminded the commission of its global responsibility. It had to show the way to a fast and just exit from coal. We reminded them that our future was in their hands.

After we left the room, the coal commission proposed to use coal for another nineteen years. Which meant that this dirtiest form of energy production was to continue until 2038. That same evening, the *Tagesthemen* (public TV evening news program) reported about thousands of children skipping school. The minister of culture commented that this must not turn into an ongoing phenomenon.[1]

Back to the Berlinale. We had been invited there only a few days after our big strike. Apparently, though we hadn't been able to move the decision makers in our country's energy politics, a group of Australian filmmakers, which was interested in the good life of the future, had taken notice. The Berlinale was to show the premiere of their film *2040*. We gratefully accepted their invitation, since it promised a pleasant distraction in the increasingly hectic daily life of the movement.

The film *2040* shows what the world could look like in 2040 if today's ideas about justice, happiness, and climate protection were to be realized. For the premiere, the filmmakers had made an unusual decision: They handed two hundred free tickets to those—young people, like myself—who would still have half their lives ahead of them in 2040.

The film takes feasible technological, ecological, and economic innovations, scales them up, and describes their best possible deployment. The result is a colorful picture of a future full of opportunities. A picture of what Germany and the world could look like if we spent the next twenty years building a sustainable future based on renewable, decentralized, and cooperative energy production—rather than on coal power for nineteen of these twenty years.

When the lights come back on, the director and I are asked to the front for a conversation. Cameras are clicking, flowers are being handed out. We must look pretty small in front of the large red curtain, at least I feel that way.

The audience is asked, too, what they think of the film, and whether they are now looking with more optimism to the future.

A small boy gets up from his seat. He says he is in seventh grade. Math, he says, is his favorite subject. One day he would like to become a shipbuilder. He clutches the microphone with both hands, directs his gaze at the director and says with a clear voice: "You know, that looks nice. But it's a fantasy. I believe it will all simply get worse and worse. I am scared."

With my eyes narrowed I look around the movie theater. Nothing moves. What you can see is young people in far-too-broad seats nodding in agreement.

I'm next up at the mic, but I don't quite know what to say. I look into the large eyes of two hundred schoolchildren and would like to spread hope, but it's not that simple. On the contrary, I know that they may be right.

Today, it is 100 seconds to twelve on the Doomsday Clock. Since 1947, the scientists of the *Bulletin of the Atomic Scientists* in the United States have been using this model clock to illustrate their assessment of the danger of a human-made apocalypse. The closest to twelve until now was in 1953, when the Soviet Union had just carried out successful tests with thermonuclear explosive devices and thus put itself in a position, like the United States, to be able to wipe out the existence of entire nations at the push of a button.

In other words, scientists consider the danger of human-made apocalypse just as great today as it was when the superpowers of the Cold War were first able to annihilate mankind from one moment to the next. But the reason for the researchers' alarm is not only the threat of nuclear war, which has increased since 2017 due to the expansion of the North Korean nuclear weapons program. The increasingly extreme consequences of global warming are another reason.

The big difference between the danger today and that of 1953 is that the danger then could "simply" be averted through diplomacy. Today the situation is different. Many global ecosystems are already so badly damaged that it may be impossible to ever fully "repair" them. And the global economy is designed to ensure that this destruction continues. That means we cannot rely on diplomacy alone. It is only one of very many steps that must be taken to avert the danger.

People like us, who still have so much of our lives ahead of us, cannot ignore this peril. Wherever we look, terrifying images dominate the scenarios of our future. We think of the year 2050 and realize that the anticipation of a fulfilled life in an intact environment is overshadowed by worry and fear: worry about a future in which conflicts over resources and poverty will dominate our everyday lives, our fellow human beings, and our view of the world. Fear of what will become of societies competing for dwindling resources. In fact, for many people this is already the reality today.

Even if all climate protection targets of all governments worldwide were met—Germany had officially abandoned its 2020 targets already in 2018—we would still be headed toward global average temperature increases of 2.7 to 3 degrees Celsius by 2100.[2] If that happens, several studies estimate, large parts of the Amazon rainforest could turn into savannah, with hardly any trees left.[3] We read of a future without coral reefs,[4] of storm-ravaged coastal cities,[5] of unprecedented species extinctions—of a planet increasingly uninhabitable.[6]

Despite the fact that we are facing one of the greatest threats humanity has ever faced, many are experiencing this drama only as a gradual process, if they notice it at all. This threat is not an asteroid hurtling toward earth. Since the consequences of our actions only become visible much later or in distant places, many people can hardly imagine the apocalyptic scale of the catastrophe. In Germany, the 2018 summer heat wave was one of the rare moments when the climate crisis was felt strongly enough in everyday life that the danger as such seemed to arrive in the public consciousness. As images of withered fields, burning forests, and dark red weather maps dominated the news, it dawned on a growing number of people how much the climate crisis was already affecting their lives, and what it could mean if this trend were not stopped. 2019 saw new heat records in Germany yet again.

THE FUTURE IS NO LONGER A PROMISE

The future—that used to be a promise. We are in our twenties (Luisa) and thirties (Alex). When our parents tell us about this period of their lives, they tell wild stories. Having grown up in a well-situated household in a relatively rich society in the Global North, they would tell us about how they grew out their hair, traveled to the North Sea or to California, went to concerts, experimented with drugs. In their first car or on the flight they could afford, they headed toward freedom.

Yes, back then big problems existed for them, too—the nuclear threat, the oil crisis, the Vietnam War, environmental destruction, and the discrimination against women and marginalized communities. But optimism prevailed. The young people of that time—and of wealthy societies—believed in the possibility of overcoming these problems. They believed in a future in which dreams of a better world would come true, injustices would shrink, and wealth unfold across the globe. If not for everyone, then at least for most of them.

We have not experienced such liberating years in our own twenties up to now, nor do we see them coming. On the contrary. The future holds no promise for us, and we know for many people it never did. Our lives are preoccupied by worries. We are not rebelling against our parents. We rather feel that we have to educate our parents' generation, who have become all too irresponsible in the course of their rebellion. We have to explain to them that their—and our— lifestyle is no longer affordable, and it never really was. We have to make them stop getting lost in the politics of the day and in their personal daily lives and start thinking about the future, taking a perspective that goes further than the education of their children. We have to explain to them that they have not looked after the world well enough. Make it clear to them that their hope that their children will be better off than they are will be dashed unless they become part of a radical change. They must understand that if they don't wake up now it will be too late.

OUR LIVES IN A MULTI-OPTIONAL WORLD

Whether we are going to the supermarket, discussing our career choices, searching online, or planning the next trip: Anything seems possible.

We have more choices and products available to us than any other generation before us. We can shop around the clock. To the "economic miracle generation,"[7] the growing variety may have felt like an improvement in options and quality of life. But already the next generation, our parents, took to the streets against the side effects of that unleashed consumer society.

It is true that the awareness around sustainability, the environment, clean energy, the conditions of the production of consumer goods, etc. has grown, but on a larger scale nothing has changed since then. On the contrary—the range of goods on offer has become more and more expansive, the variety more and more absurd.

We stand in front of air-conditioned supermarket shelves and have to choose between the shrink-wrapped organic cucumber from Spain and the conventional cucumber from the region. Or the cheapest, the conventional brand from France. At a time when plastics are polluting the world's oceans, industrial agriculture ruins the soils, and transport across long distances increases the carbon footprint of products, none of these options is compatible with the goal to leave an intact planet for future generations. We are placed before a choice that isn't one. People tell us that this is the essence of modernity and freedom. The truth is that we are imprisoned, knowing full well that in each case we are deciding against our own future.

The same is true for our clothing, for technical appliances, furniture, household goods, for means of transportation and communication, and the houses we live in. Rarely do we have the choice of living in harmony with nature and of acting responsibly in view of the generations that will follow us.

WE ARE PART OF THE PROBLEM

No generation before us was granted as many privileges as we have in Central Europe today. For middle-class children like ourselves it is easier than ever to gain experience abroad while still in school. A language exchange with France, a vacation project in Tanzania, a class trip to Sweden, and then a school year in the US, Australia, or Chile—these are normal today.

We belong to the privileged 18 percent[8] of the world's population who have been on an airplane. That is about to change, however, because never before has it been so cheap to buy a ticket, never before have so many people flown, and every year more and more are doing so. But whereas flying was a symbol of freedom and quality of life for our parents' generation, for us every mile we fly comes with the unpleasant awareness that we are contributing to ocean acidification, species extinction, and global warming. There is even a word for it now: Flight shaming.

In the foreseeable future, the distances individuals can and may fly will have to be severely restricted if we truly want to limit warming to 1.5 degrees or at least 2 degrees. The technological developments in climate-friendly air traffic are too slow, the current climate damages already too big. Thanks to the many opportunities for travel already during our school years and the communications channels provided by the internet, we have friends in many parts of the world. But to live out this world citizenship in good conscience will be impossible. Therefore, it is anything but liberating: We have more and more opportunities, but if we use them we also have to live with the realization that we are destroying the basis of life for our descendants and those living in the most affected areas today.

This dilemma also concerns the question whether we want to become parents one day. Is it the responsible thing to do, to bring children into the world, when according to all predictions this world will be shaken by unimaginable crises? And: Is having children responsible toward our fellow humans, since statistically nothing leaves a larger CO_2 footprint than a child in industrial societies like ours?

It is a frustrating experience: The attempt to embrace a "future fit" lifestyle in this situation is inevitably doomed to fail.

We want to do things right: We save money for train tickets in order to avoid flying, but airline sales continue to skyrocket. We reduce animal products in our diets, but factory farming keeps getting subsidized. We switch to our bikes, but more and more new diesel SUVs are being sold elsewhere. Since the societal structure that surrounds us is not sustainable, our individual uprising against it remains a battle against windmills. "There is no right life in the wrong one," Adorno said. In the context of the climate crisis this sentence hits the nail on the head: There is no sustainable life in an unsustainable society. So emissions keep rising.

This knowledge is the burden we carry around with us. In Germany, we have grown up with it: with recycling, energy-saving lamps, water-saving showerheads, and the habit of printing on both sides. We never understood the note under emails saying, "Please remember the environment before you print this email"—we would never have thought of printing an email. And who even owns a printer nowadays? Still, or perhaps precisely because of this, every shopping trip, every vacation, and every purchase decision that we make is combined with the feeling that we are living at the expense of others.

It is not about "growing up"—as the sociologist Cornelia Koppetsch condescendingly remarked in a lecture about Gen Y[9]—but about the awareness of the dilemma of growing up in a society that dumps every societal responsibility on the individual.

People, no matter what age, should not have to constantly be faced with a decision between products and services that are rarely compatible with human rights and ecological standards, or those that protect the basis of life for future generations. It cannot, should not, must not be up to the individual to have to decide for or against the future in every single thing we do.

We want to do everything right—and can't. We are ourselves part of the problem. We are overwhelmed. Overwhelmed by the ostensible magnificence of this multi-optional consumer society. And overwhelmed with the understanding that someone will one day have to pay for it: and that someone is us.

NAURU: THE CANARY IN THE COAL MINE

Seen from Central Europe, Nauru lies literally at the other end of the world, in the central Pacific. By boat it's a multiday trip from there to the neighboring islands of Kiribati, Tuvalu, the Marshall or Solomon islands: 3,500 kilometers from Australia's Brisbane, around 200 kilometers from the coast of Papua-New Guinea.

With barely 10,000 inhabitants on 8.1 square miles, Nauru is one of the world's smallest republics, but not only that: It is also the site of sad superlatives. For instance, 77 percent of people here suffer from obesity, which is a world record. Coronary heart disease and kidney failure are statistically most prevalent here, and almost one in three people suffer from diabetes.[10]

How could it have come to this? In the 1970s, Nauru was one of the richest countries in the world.[11] The reason for this wealth was huge phosphate deposits, which had formed over millennia from the droppings of migratory birds that rested on Nauru. Phosphate is an indispensable ingredient of fertilizers for agriculture, and nowhere in the world is phosphate as pure as on Nauru.

For three thousand years people have lived on this remote island, far from the rest of the world. The Western narratives would call it "undiscovered" until the end of the eighteenth century. That changed in 1798, when the English captain and whale hunter John Fearn came upon this place, which became known as Pleasant Island to the British public from his reports about its peaceful population. At this time twelve tribes were living on Nauru. They had divided up the area between them and lived above all from coconuts and fishing. It wasn't until the British Empire expanded its sphere of influence in the Pacific in the middle of the nineteenth century that Nauru, too, was pulled into the maelstrom of imperial ambitions.

At first, prisoners from Norfolk Island arrived on English ships. They were followed by more and more people from Europe, who settled there in their search for a pleasant life in the South Sea. With them, the peaceful community slowly came apart. Some tried to seize power over the island by force; others promoted the trade in copra, the dried fruit of the coconut, which was the local population's most important sustenance. As trade expanded, so did the interest of European colonial powers, especially Germany and England. When at the end of the nineteenth century they divided up their spheres of influence in the Pacific, Nauru became a German protectorate.

At this time the workers of the Pacific Islands Company from the Australian city of Sydney discovered the phosphate deposits that would later form the foundation of the island's rise and fall. In 1907 the British Pacific Phosphate Company became the first company to mine phosphate on Nauru. They paid annual taxes to the German government and a negligibly small fee to the Chinese guest workers in the mines.

With the outbreak of World War I Nauru became a geopolitical pawn of the superpowers. For a short period, Australian forces occupied the island. During the postwar decades it was under the influence of Great Britain, which shared the phosphate profits with Australia and New Zealand. During World War II, Japan occupied the island for military purposes, US bombers flew attacks on Nauru, and a famine broke out. It was not until January 1966, almost sixty years after the start of phosphate mining, that Nauru gained independence, and within a few years it became one of the richest countries on the planet. The founding president Hammer DeRoburt nationalized the phosphate mines, and from that moment the profits of the newly founded Nauru Phosphate Corporation flowed directly into government coffers. With the Nauru Phosphate Royalties Trust some of the income was invested abroad, because it was clear from the beginning that the phosphate would only last for thirty years. The money flowed into luxury hotels and estates, office buildings, and shopping malls in Australia and the US.

People from the surrounding islands carried out the arduous work in the mines, while the former guest workers from China opened up shops, restaurants, and grocery stores in order to profit from the new economic boom. The Naururians, however, were swimming in money. Through real estate deeds and titles, for the use of which the state was paying them, many became rich very quickly. Although the state did not collect taxes, it financed schools, health care, electricity, overseas study programs, and at one point even private house cleaners and cinema tickets. The people of Nauru traveled around the world and brought back the newest television sets and stereo systems from Europe, the US, and Asia. They kept expanding their houses, and ever-bigger pickup trucks were shipped to the island in large containers and driven around in circles on the island's only street. Groceries were imported from Australia so people ate the fruits of the fields that had been fertilized with Nauruan phosphate. Nothing was produced any more, nothing was repaired. What broke was disposed of and bought new. It is said that there were parties on Nauru where the guests used dollar bills as toilet paper. In short: The sudden wealth catapulted the people of Nauru into an increasingly grotesque consumption frenzy. And it led Nauru straight to its ruin.

When the phosphate price dropped at the end of the 1970s, the island began to feel its desperate dependence for the first time. The production volume dropped, reserves disappeared. At the same time, foreign investments did not yield any returns, money kept disappearing. But when the price stabilized, this warning was quickly forgotten.

This phase of stability lasted until 1986. President Hammer DeRoburt was in power almost the entire time and kept the country on its growth course. Turbulent years of government turnover followed, with twenty-two presidents in twenty-three years. To maintain the country's lifestyle through shortages, the government took out loans in the 1980s and 1990s. When the phosphate extraction was reduced to a minimum in 1997, older debts were paid with new loans.

In its desperate search for new sources of income, the government got caught in an increasingly fast consumption-driven downward spiral. Nauru attracted hundreds of offshore companies and became one of the largest tax havens in the world. The Russian mafia alone allegedly laundered 70 billion dollars through shell companies in Nauru. More and more cases of corruption became known, including those in which the government and its advisers embezzled money. People said the government made 7.4 million dollars by selling passports, which were found among terrorists who sympathized with Al-Qaeda, among others.

Then at the close of the 1990s came the end: The Bank of Nauru went bankrupt, and many inhabitants lost their entire savings.

Under pressure from the debt burden, the government of Nauru sacrificed its last ethical standards. When the Australian government refused entry to a boat with Afghan refugees in 2001, Nauru accepted something like a Pacific version of the EU-Turkey Deal: A camp for the asylum seekers was erected, and the island received 30 million dollars annually to run it. Up to 1,200 refugees have since been housed there under miserable conditions. Human rights organizations keep reporting bodily and sexual abuses, self-immolation, lack of health care, and the catastrophic situation of children.[12] In early 2019, there were still 350 asylum seekers imprisoned on Nauru; in 2021 the contract with Australia was extended to keep asylum seekers there indefinitely.[13]

The history of Nauru should be a warning to us all, just like the famous canary in the coal mine, which stops singing before we run out of air. We could learn from this story what will be in store for us if we don't fundamentally change our relationship with nature. Of course, Nauru is special, an extreme case. But the straightforward and isolated case of this island can show us more plainly than anything else what the consequences of living at the expense of future generations and nature can be.

There is another reason why Nauru is important for understanding the climate crisis. Islands like these are especially threatened by sea-level rise and the increase in storm surges. The inhabitants of Nauru live almost exclusively along the coast. As a consequence of the many years of phosphate mining, a large part of the island's surface is an uninhabitable moonscape. The highest point on the island is only sixty meters above sea level. With every meter that the water around Nauru rises, parts of the population are losing their homes.

According to the diagnosis of the UN Development Programme, climate change endangers food security and public health on Nauru. The marine ecosystems are threatened by sea-level rise, higher water temperatures, and increasing numbers of storms. Finally, coastal erosion and highly stressed coral reefs are threatening fishery yields, now one of the most important sources of food on Nauru.[14]

The story of Nauru resembles a tragic parable grounded in Western imperialism about the consequences of human greed and short-sightedness in the capitalist economic system. It tells us how a society blinded by wealth destroys the foundation of its own existence within a few decades—because, comfortable as it had become, it did not want to accept that this intoxication based on exploitation of natural resources would ever come to an end. The history of Nauru also tells of colonial exploitation, geopolitical power struggles, and individual enrichment at the expense of the common good.

2 BECAUSE YOU ARE STEALING OUR FUTURE

LUISA From the Invalidenpark it takes roughly an hour until you reach the Bundestag in Berlin. At least when you are moving as part of a demonstration. In between, we keep stopping so the photographers running ahead can take their pictures—of the front banner and the long row of signs and faces.

Shortly after our march has begun to move forward, we are walking under the bridge that connects two parts of the Charité hospital building. Up there, people in white coats are taking pictures from behind their windows and waving at us. We wave back. The climate crisis is also a health crisis: More and more heat waves endanger young, old, and sick people. Ever-warmer average temperatures lead to tropical diseases spreading in Europe. For more and more people the climate crisis is becoming a psychological burden. In short: The climate crisis is making us sick, and that's why we see the hospital employees as our allies.

After the Charité we walk straight down *Luisenstrasse*, past the offices of various NGOs, where people lean out of their windows and take photos of the bird's-eye view, which they tweet out later on. A group of pharmacists is crowding behind a shop window, too. Then comes the commuter rail bridge, directly in front of the Spree River. When we reach the bridge, two people are standing left and right, close to the wall. They were already expecting us. With narrowed eyes they count the passing participants. With an average shoulder width of sixty centimeters almost exactly twelve people can walk side by side under this bridge; it is the optimal place for an accurate count.

Normally, about two hours after the beginning of a demonstration, the news agency dpa (Deutsche Presse-Agentur) calls and asks us about the number of participants for their agency report. At this point, if everything goes smoothly, the last ones have just passed the bridge. The numbers are sent out into the world. Although those numbers cannot convey the atmosphere or any meaning, they do make it possible to grasp what we are carrying into the streets. At the first strike in Berlin there were 350 people. Four months later there were 25,000.

There are a few adults among the demonstrators, but the majority are children and teenagers. Entire preschool groups and elementary school classes have come with their teachers. They wear shirts in the same color so nobody gets lost in the crowd. One elementary school block has dozens of classes combined. Many have made their own signs. One girl is holding a small cardboard sign up into the air, and she has written with a Sharpie: "When we said shut it down we did not mean your brain." A father has come in his son's place. "My son Willi is only turning two, but I'm here," his sign says, which he holds high above his head. A young man carries the slogan: "If the earth were a bank you would have saved it already." In the middle of it all are the TV crews, directing their cameras at the children and their signs, trying to capture the atmosphere.

Alongside some who have been protesting for environmental protection for years walk thousands of young people who are doing it for the first time. The euphoria of so many people who realize what it's like to be part of something big mixes with the fear of the masses, the unknown. You don't learn how to protest in school; you learn it here, step by step.

Shortly before we stop the demonstration at the chancellor's office, we march in between two high government buildings. Next to me are the familiar faces of Carla, Jakob, Linus, Louis, and all the others. We get ready. The shade of the tall buildings darkens the street. From the windows above, people in suits are looking down on us. This is where, in the very epicenter of the Federal Republic, we turn on our megaphones.

"We are here, we are loud, because you are stealing our future!" we shout together. Again and again, rhythmically, until we are hoarse, on and on. Thousands of voices, of four-year-olds and twenty-four-year-olds, all in time, echoing upward between the buildings. Sometimes we stretch out our arms into the air. Goosebumps. We chant our slogan at every strike, no matter whether we are ten or thousands, with banners or microphones, in chorus, at times in anger, at times less so, but always to the point. Because as simple as it sounds, it really gets to the core of the issue: Something is being stolen from us, taken from us, withheld. Something we have a right to: our future.

At the annual shareholder meeting of RWE, the energy company operating the dirtiest coal power plants on the continent, I formulated it this way:

"We are coming of age in a world in which climate chaos is becoming the norm. In which humanitarian catastrophes, climate migration, supply bottlenecks, and species extinction will dominate our lives. In which a self-determined life, our future, is overshadowed by the collapse of the ecosystems around us. That is what the science predicts."

Parts of my speech were met with boos. A few people applauded. When I left the hall later, I felt as though I had invaded foreign territory. Rarely have I met with so many so hostile stares. But there were exceptions. As I picked up my backpack at the coat check, an elderly woman turned around and whispered in Rhenish dialect: "I have been coming here for twenty years. Back then I protested against nuclear waste transports. It was high time someone started a fire under these guys."

It is not a novelty that young people stand up for their future. Or that they accuse older generations of living at the expense of the young. We are not reinventing the wheel of youthful indignation. But the commitment to our future has a new quality today. In the last ten to twenty years some important starting points have been reversed.

A SCIENTIFICALLY FOUNDED FEAR OF THE FUTURE

Traditionally, fear of the future is the result of a measure of uncertainty. That is the ghost train effect of the discourse about the future—if ghost trains were brightly lit, we would not be spooked. We get spooked because we cannot see everything, because we do not know what's coming.

After World War II, it was the Cold War that allowed for the threat potential to skyrocket: The aforementioned Doomsday Clock was an instrument meant to indicate the risk of apocalyptic nuclear war. One way the fears, which were stoked by the great uncertainty about what tomorrow would bring, were rendered visible was in the catastrophe film. In December 1983, the American television film *The Day After* came to German cinemas. It depicts the catastrophic consequences of a fictitious nuclear war between the United States and the Soviet Union. In the film, the nuclear conflict destroys the infrastructure in the US. We see uninhabitable cities, a collapsed civilization.

But even long after George Orwell had written his novel *1984*, the darkest visions of the future, while inspired by experiences of the past and present, were always science-fiction stories. Though they were not improbable, they only described possibilities, not certainties.

For us this situation has changed. Whereas the danger used to emerge from the unknown, it is now the known that makes us shudder—and drives us onto the streets. Today, what will happen under certain conditions can largely be documented scientifically. For several decades now, future scenarios have been calculable like never before, at least with regards to the fate of natural habitats. Year after year, technological and mathematical feats allow for more precise forecasts.

Since the year 2000, the reports of the Intergovernmental Panel on Climate Change have regularly presented different scenarios, which outline where we are headed, depending on varying factors such as the extent of globalization, technological progress, and population development. Eighty percent of young people in Germany are rightfully worried when they think about the environmental conditions under which future generations will have to live.[1]

Because even in moderately pessimistic projections the conditions are catastrophic.

LUISA Together with Greta Thunberg, I took a look at the supercomputer located in the basement of the Potsdam Institute for Climate Impact Research (PIK). It can determine with unprecedented accuracy how the climate systems will develop in thirty, fifty, a hundred years, based on the emissions expected in these periods. This creates an oppressively accurate picture of the natural conditions under which we are likely to grow old.

When Greta and I visited PIK, five of the country's leading climate researchers spent two hours with us. The researchers showed us around the institute, which is situated on a hill in the middle of a forest south of Potsdam. This is also the site of the famous Einstein Tower, which was used to empirically confirm Einstein's theory of relativity. From the outside, the idyllic location of the institute is more reminiscent of a health resort than the center of modern climate research. The researchers showed us their steel-jacketed equipment, high-performance computers humming away in a cellar deep below the institute, protected by cages. There is very little oxygen in that musty-smelling basement, so you can't stay there for long. Later, in their wood-paneled office, the scientists explained the situation to us like this: Emissions over the last 140 years have turned humanity into a geological force. We are creating deserts, we are changing the way oceans and air masses circulate, we are destroying glaciers, and we are terrorizing in extreme ways the very ecosystems on which we depend. Just a few decades ago, these experts told us, some of them could not have imagined how damaged the planet would already be by 2019. I had to swallow hard when I heard that. At that moment, I wished that all those who are always accusing us of scaremongering could have heard it in exactly the same way.

Many scientists are convinced that human impact on nature has by now become so disruptive and destructive that they have given the current geochronological epoch a new name: Anthropocene. The Age of Man. This means that human beings are now considered the greatest influence on global biological, geological, and atmospheric processes.[2]

In terms of human history, this realization marks an epochal change for the relationship between humans and nature. For the first time, people are unhinging planetary processes without thinking about tomorrow, not to mention the day after tomorrow. On the contrary: The overwhelming majority of serious climate science predicts short- and long-term environmental changes beyond anything humans have experienced so far, and, at worst, can survive.[3]

THIS CRISIS COULD HAVE BEEN PREVENTED

The current climate crisis did not happen by accident. The facts have long been known. But because of the neglectful handling of this problem, the danger is now gigantic and the probability of overcoming it successfully has become minimal.

For almost half a century, science has been able to calculate the consequences of human climate destruction—there are books that tell the story of when which environmental and climate changes were first predicted.

A pioneer in this field was the Swedish chemist Svante Arrhenius: He already determined in 1896 that the burning of fossil fuels could lead to global warming. It's maddening, yes: The first indication of the developments that we are experiencing today is already 120 years old! In 1969 a distant descendant of Svante Arrhenius was born. His parents decided to name their child after the chemist in the family who once recognized the connection between emissions and global warming. That seems to have been a broad hint from history—because today this Svante is the father of the world's most well-known climate activist: Greta Thunberg.

In 1958, climate scientist Charles Keeling began the first continuous measurements of CO_2 concentrations in the atmosphere. The measurement curve became known as the Keeling Curve.[4] Keeling was the first scientist able to show that CO_2 concentration increases due in part to the burning of fossil fuels. His curve captures how the concentration of the greenhouse gas CO_2 decreases in the northern hemisphere in the spring and increases again in the fall—which results from the increase in CO_2 reduction due to spring vegetation growth. Subsequently, as early as the 1960s, oil and gas corporations like Exxon-Mobil commissioned scientists to investigate the causes and consequences of global warming. They wanted to know what was coming. After all, they were the largest emitters of CO_2 on earth. From their data collections emerged a whole host of studies about the potential of global warming. Since 1966, so-called TIROS satellites have made it possible to measure the intensity of solar radiation and the earth's heat balance. In the 1970s, a number of climate scientists predicted that the warming caused by the emissions at that time could lead to an ice-free Arctic by the year 2050. Since then, scientists have been exposing and explaining the causes and consequences of human-made climate change piece by piece.

LUISA "At the time it seemed quite simple," Bill McKibben told me a few years ago on a Swedish train that itself seemed to date back to the 1970s.

Bill McKibben was among the first to bring knowledge of global warming to a wider public. His popular science book *The End of Nature* came out in 1989. A year earlier, climate scientist James E. Hansen had testified in Congress that 99 percent of global warming was not a natural development, but was due to anthropogenic, i.e., human-made greenhouse gases. This landed Hansen on the front page of the *New York Times*.[5] At that time McKibben was also reporting on the climate crisis as a journalist and author. He was irritated by the lack of public awareness of the explosive nature of the findings. That's why he began to explain the facts in his books. In 2007, together with students from his seminar, he co-founded 350.org, today the world's largest nongovernmental organization devoted solely to the climate crisis.

I met Bill in 2014 at the award ceremony of the Alternative Nobel Prize in Stockholm. Back then I had just graduated from high school and was working in the editorial department of an environmental magazine, which had given me the task to interview Bill.

I was impressed right away with the patience and composure with which he dealt with climate deniers—people who are dominating the climate discourse in the United States in unparalleled ways. And by the clear language he found for the big problems. About German climate policy he said: "Germany brought the world so much grief in the twentieth century—in the twenty-first it would have the opportunity to bring the world great joy."

While the frozen Swedish landscape passed by outside the train window, Bill told me that back then he had wanted to write a powerful, comprehensible, and scientifically accurate book about climate change, which would explain what was going on with our planet—so people would finally realize the seriousness of the issue and politicians would act.

Well . . . *The End of Nature*, as mentioned above, appeared in 1989 and it did indeed become a bestseller. It was translated into twenty languages. Millions of Americans have read the book. Even in a country where a third of the population still doubts that human-made climate change exists,[6] it has, step by step, moved into public consciousness. But politically nothing has happened—on the contrary. The US experienced a growth of climate "skeptics," the coal and oil industry boomed, and per capita CO_2 emissions remained the highest in the world for another twenty years.[7]

What happened with Bill McKibben's book is exemplary. It's not as though one couldn't know what our way of life, our way of doing business and growing, was going to mean for future generations.

In Germany, *Der Spiegel* ran a cover story about "The Climate Catastrophe"—that was in August 1986. Their story begins with a future scenario: "The catastrophe had not come as a surprise. Scientists had warned in good time, environmentalists had demonstrated tirelessly. Finally, even politicians had recognized the seriousness of the situation—too late: It was no longer possible to prevent the worldwide climate disaster. Now, in the summer of 2040, New York's skyscrapers rise like reefs from the sea, far off the coast. Hamburg and Hongkong, London, Cairo, Copenhagen, and Rome, have long been flooded, swallowed by the sea."

To fathom the extent of political failure, it helps to look at the United States. At the end of the 1980s, 68 percent of the US population knew about the phenomenon of global warming, and one third of citizens even said that they were very concerned about it.[8] The topic was in vogue. Republican George H. W. Bush entered the presidential campaign as a self-declared environmentalist and said he was determined to tackle the greenhouse effect with "the White House Effect." Bush never fulfilled this promise, but the green credentials gave him tailwinds on his way to the White House. Years of political discussions, publications of scientific testimony, media reports, and hearings in Congress predated this peak in public attention.

After forty-one members of Congress had demanded that President Ronald Reagan initiate an international treaty along the lines of the Montreal Agreement on ozone, he signed a declaration that promised cooperation on the containment of global warming together with Mikhail Gorbachev, then general secretary of the Communist Party in the Soviet Union. Back then, too, a record-breaking heat wave had increased public attention in America: In Alaska and the western parts of the country dozens of wildfires were burning, rivers evaporated, the Mississippi shrank to one fifth of its usual volume in some places, and in New York streets began to melt in the sun. Some days saw no temperatures below 38 degrees Celsius (100.4 degrees Fahrenheit) anywhere.

As early as 1990, activists, engaged politicians, and responsible scientists had closed ranks—similar to what we are seeing today with the organization Scientists for Future—which finally produced the political momentum in which the signing of an international climate protection agreement under the leadership of the US seemed within reach.

In 1988, the Enquete Commission presented its three-hundred-page interim report "Precautionary Measures for the Protection of the Earth's Atmosphere" to the German parliament. In its introduction, the commission emphasized the "urgency" and the "global significance of the task at hand" and warned that "the scientific findings on the greenhouse effect . . . are already so compelling in their basic statements" that "far-reaching measures to reduce greenhouse gas emissions must be initiated as quickly as possible." They pleaded with parliament to debate and vote on their recommendations for action "as soon as possible."[9] Nearly everything we say in this book about the scientific knowledge of the climate crisis was already in that report: about the anticipated development of emissions, their catastrophic consequences, and even possible solutions to prevent the present crisis.

The facts have been public knowledge since the 1980s, or at least since the 1990s.

Since then, however, over the past three decades, humans have emitted more CO_2 than in the entire history of mankind beforehand. This means that mankind has knowingly done more damage to the planet than unknowingly in the preceding millennia.

Since those who will have to deal with this situation in one way or another all their lives are unfortunately the youngest in our societies, older generations are being collectively taken to account. They are liable for leaving us a natural environment in shambles, for not acting when there was plenty of time to do so, and for not stopping when it was already clear where endless growth and endless resource exploitation would lead.

NOT A BRAVE NEW WORLD AS WE LIKE IT

It's about sticking to agreements that were made a long time ago, in every sense of the word. The Paris Agreement, which was adopted on December 12, 2015 is thirty-two pages long. Thirty-two pages, divided into twenty-nine articles, stating what needs to be done.[10] The text is the result of decades of negotiations, a milestone of diplomacy, which was also unanimously approved by the German parliament half a year later.

The agreement lists the important points. It's true, they are written in convoluted and rather abstract UN language, and not especially broken down for the conditions in each country. But the basic idea is anchored here: According to this agreement, the global community will strive to reduce warming compared to the preindustrial age to an increase of 1.5 degrees, or at most "well below 2 degrees." The principle of "common but differentiated responsibilities" applies, which means it is a collective responsibility, but rich countries like Germany have an extraordinary responsibility to take ambitious action and beyond that to support poorer countries with their climate protection. "Intergenerational justice" is also mentioned, by the way. Older people bear more responsibility than younger ones.[11] Everything is there, long decided with the blessing of the German parliament—at least in theory. How can an initiative that was agreed upon through democratic process be so blatantly ignored? We demand that our parents follow the rules—at least in principle—the rules that they themselves have created: the climate targets and international agreements.

A GLOBAL QUESTION AND A GLOBALIZED GENERATION

No, Germany has not been the first country to feel the harsh consequences of the climate crisis. We live in a socioeconomic buffer zone with relatively high resilience vis-à-vis environmental dangers. This doesn't mean the climate crisis has no impact on Germany—in summer 2021 more than 150 people died in record-breaking floods in the west of the country. Scientists from the University of Oxford later stated that the floods were up to nine times more likely due to the climate crisis.[12] And still the climate crisis is usually portrayed as something that happens in the far distance or far future. But this starting position is becoming less important. If you tell young people in Germany that they should go see the world, get to know other cultures, feel at home in Europe or better yet in the whole wide world—hence if you educate what is likely the first global generation in history—you must not be surprised that so many develop a responsibility and a consciousness for problems that arise outside our country's borders. The call for honest climate protection has a globalized core. We are not doing this just for ourselves, but also in support of and solidarity with those who are growing up less privileged. In a world that is growing closer together every day, the problems of others are increasingly becoming our own.

Since nowadays so many young people take participation in international exchanges for granted, we understand that our responsibility is also an international one. That it is not (only) a question of what percentage of global greenhouse gases a German coal-fired power plant emits. It is also about responding to the cry for help from a generation that is represented on every continent. The climate generation, if that's what you want to call it.

HUMANITY HAS A DEADLINE

Humanity's room for maneuver is already very small. And it is shrinking rapidly. In the summer of 2019 we had put it at 350 gigatons. In 2022 it was down to 308 gigatons. 308 gigatons of emissions that can still be emitted to keep global warming at 1.5 degrees Celsius.[13] If this budget is exceeded, the world will accumulate damages that cannot be justified. Hundreds of millions would lose their livelihoods, and a multitude of ecosystems would collapse. At the rate the global community is currently emitting, this emissions budget will be reached in less than seven years. Which means we have less than a decade left—that's the blink of an eye in world history. And every second that passes, another 1,331 tons of our emissions end up in the atmosphere, turning up the dial of global heat, bit by bit.[14]

When we claim today that our future and the present of those living in the most affected places is being stolen, we do so on the basis of an urgency hardly ever seen before. The time pressure has become unbearable. No, it is not about solving all problems here and now. Or, as a few right-wingers polemicize on Twitter, to immediately "shut down all coal plants and discard all automobiles" everywhere. It is about taking the path now that leads to the 1.5-degree target. To achieve this, the rise in global CO_2 emissions must be stopped and reversed to an annual reduction within the next few years. Above all, this is the responsibility of the largest emitters, like Germany. Based on the size of its population, Germany has a carbon budget of 3 to 6 gigatons left as of 2022. That's the total amount of CO_2 that we are still allowed to burn so that the 1.5-degree goal can be met with a certain degree of probability. In order to stay within that budget, emissions must drop sharply so that they are at zero when the budget is used up. But given Germany's current emissions path, the budget will be used up around 2030—and the emissions will still be very high. And every year that we cut emissions less than necessary, the gap between the actual and the target state will continue to increase. Year after year, the measures to get us back onto the path toward Paris would have to be more radical. Until a time when you would literally have to say that people must abandon their cars in the street. That's what we are trying to avoid.

Climate activists are often said to be "impatient" or "frantic." That impression is understandable.

In the case of the climate crisis, however, the urgency is dictated by geophysics. Physics dictates the speed at which we must act. This is one reason why tens of thousands of scientists have joined together as Scientists For Future in order to support our demands. They look at the climate graphs and see that certain climate tipping points have occurred much faster than expected, and that the window of opportunity that we must use is closing fast.

WHO IS STEALING OUR FUTURE?

LUISA When I met Angela Merkel, it was so hot that I asked myself how she could bear it in her thick red blazer. We were in Goslar, a small city near Göttingen, where her party, the CDU, was founded almost seventy years earlier.

For as long as I have been able to think politically, Mrs. Merkel had been the German Chancellor.[15] Her first election in 2005 was one of the first political events I consciously followed, and like probably most during the years of her government, I had never wondered what I would say to her if we ever met.

Prior to our meeting in Goslar, I and several others had already spent over two hours discussing climate protection with French President Emmanuel Macron, Barack Obama, and five German federal ministers. On the occasion of the chancellor's visit, the Fridays For Future group in Goslar had organized a climate strike, although it was a Wednesday. They also asked me to write a speech. And although we had spoken with several heads of state before, this meeting felt particularly important. I had thought long and hard about what I would tell the chancellor. I had hoped that it could become a good conversation.

In June 2019 it was already foreseeable that Mrs. Merkel would not be in office for very much longer, and I expected her to be thinking about her legacy. My speech was entitled "Become our Ally!"

One part went like this: "You are still one of the most powerful women in the world. Do something with it while you still can. Become an ally to the pioneers, the world leaders, the movers and shakers. Become an ally to those of whom one will say, years from now, that they acted when it was still possible. An ally to those who understand that climate protection is non-negotiable and that there can be no 'measured balance in the middle'—climate protection cannot be 'measured' but must always be radical. That's what this crisis demands, radicalism in the best sense.

"Become an ally to those who don't accept excuses, who show the old guys afraid of change how it is done, those who understand that physics cannot be appeased with words, and those who are ready to make unpopular decisions, because all else would lead to catastrophe.

"What is stopping you? Become an ally to those of us who are rolling up our sleeves now. We invite you. What could possibly be stopping you?"

Perhaps, I thought, there was a tiny chance that the chancellor would go for it once more in her remaining days. Really go for it, without allowing the nagging cabinet or the hesitant or the populists to hold her back. I had that hope. But apparently, our generation's call did not reach the chancellor's office.

When it was Angela Merkel's turn to speak, she said things like, "It's okay for climate protection to take one more day." She said that across the barrier into the agitated crowd of about a hundred climate strikers. Nobody felt sure whether to listen respectfully or demonstrate loudly, which resulted in a confused mix of the two. "We are already doing things," Merkel added. And muttered something about "commitment," and that she liked our slogan "No More *Pillepalle*" (No More Small Steps).[16]

I was speechless. I had expected many things, but not that. Not after six months of strikes, six months of political perplexity about the adequate reaction to schoolchildren suddenly demanding climate protection. Merkel is a physicist. Shouldn't she understand what it means when climate graphs rise steeply? And that physical tipping points don't wait for governments in disarray to agree on what to do?

I actually never gave the speech that I quoted above. When, shortly after Merkel, I stepped onto the stage, I said instead: "We will initiate the greatest transformation that the world has ever seen. We will write history, about a generation that was unstoppable. I'm not sure Mrs. Merkel will be part of that history. And that's okay, too. We don't have the time to wait for every last chancellor to get that we have to start now or never."

On the way home on the train I pondered for a long time who really was on "our" side. Who "we" were anyway, and who wasn't. And how we would manage to inspire society for our cause in a way that was both challenging and inviting.

A week later, at the *Kirchentag*[17] in Dortmund, I invited the Protestant Church to become our ally. Those present cheered.

The big questions stay with us when it gets quiet again, when the demonstration signs are packed up and people, their feet frozen and their faces sunburnt, start the journey home. The crowd disperses, but the discussions continue in living rooms and also in state parliaments, offices, and corner pubs. These conversations must not be underestimated; they are an indispensable part of the movement. It's easy to shout with hundreds in the street, "We are here, we are loud, because you are stealing our future!" It's easy to speak of "us" and "them," of those who cheat and those who get cheated, as long as you think yourselves in the same camp. It's less easy to translate that slogan, which holds so much accusation and reproach, into a conversation with those on the other side.

As citizens of one of the richest countries in the world we grow up in great prosperity. How dare we young people accuse those of theft who have worked hard for all this, and who gave us a life full of choices?

Some find it arrogant that we talk about "us" and "them" on the street. They say that we, the young, have not done anything for the country and its prosperity. That we don't appreciate what generations before us have achieved. That's understandable. The older generation looks back and sees everything they have accomplished, and how rapidly the country has changed over the last thirty years. They look back on unprecedented economic development and also on many efforts for the environment and climate protection. This leads them to assume that "these young, loud, and historically oblivious people" want to wipe it all away. After they, the older people, had put so much effort into it. Shouldn't we give them a little pat on the back for that? Instead, we stand in front of their homes and offices by the thousands, protesting, because we disagree. It is a question of perspective: The older generation looks back and remembers where we came from, and they see how far we have come. We, however, look ahead toward where we still have to go, and we see that the path to an ecological future, a society that lives within its ecological boundaries and respects the Paris climate targets, lies in the seemingly unattainable distance.

It is important to differentiate somewhat here. Because the sweeping accusation of future theft hits different actors in different ways: We are collectively targeting our elders, the generations before us, because they should have acted sooner. As parts of a civil society they should have demanded effective climate protection and made their voices heard together. They failed to avert the looming catastrophe through public pressure and mobilization. They knowingly lived above their means and watched as a relatively small group of people was able to bring the climate crisis to a head for personal gain.

That group is the target of our second accusation. It is directed at the key players in politics, finance, and the economy. They are the primary drivers of this crisis. This group includes, for example, the one hundred energy firms worldwide who have been responsible for over 70 percent of global emissions since 1988. People like Ben van Beurden, the CEO of Shell, who heads his corporation with an annual salary of €9.7 million, and who is responsible for 1.7 percent of the greenhouse gases of the past thirty years. Imagine, if these hundred corporations had changed their CEOs every five years—they would fit into one single InterCityExpress (ICE) train.[18]

True, these companies were only able to flourish for so long because there were buyers for their products. And of course, these CEOs are not the only ones responsible for questions of strategic direction, the list of stakeholders is significantly longer. But every single passenger in our fictitious ICE train sat in the center of power. Every single one should have assumed their global responsibility and should have initiated a course correction in time.

These people pocket their profits or pass them on to their shareholders. They are supported by their investors and by far too many politicians who refrain from interfering in order to secure their power position with reliable connections to industry.

Yes: It is easy to label us as ungrateful. And understandable. All of a sudden, we appear on the street and complain that all the prosperity is worth nothing, that all the effort, work, time, and energy that has flown into environmental and climate movements over the past thirty years was not enough. You can certainly see it that way. But an important aspect gets forgotten: Would we even take to the streets if we hadn't learned from generations before us? If we hadn't been sensitized by all the work of our predecessors? If we had not learned from history that mass mobilization can influence political processes?

THE FIRST STEPS OF A MARATHON

Instead of forgetting history, we want to take a close look at it. Social movements, awareness-raising campaigns, international cooperation, and mass protests have led to incredible successes in the past in promoting developments that had previously seemed completely utopian. Just think of the successful fight against forest dieback in the 1980s, or the international efforts against the depletion of the ozone layer in the atmosphere, the so-called "ozone hole."

Both environmental problems are historically exemplary for effective problem solving as a result of effective awareness-raising. In those cases, political debates led to effective measures. In the case of the so-called "forest dieback" the prophecy of the death of the German forest raised society's lasting awareness of the environmental consequences of pollutants in the air. This led to a majority supporting demands for better filters on power plants, oil-fired heating systems, and exhausts. As early as 1983, far-reaching political measures were taken. This was just two years after the *Der Spiegel* article "The Forest Is Dying" had first alerted the public to the urgency of the issue. Only four years later, in 1987, the Montreal Protocol was adopted, an international treaty that is seen as a prime example of successful global environmental diplomacy to this day.

In 1974, British researchers had discovered that so-called CFCs, chemicals used in spray cans and refrigerators, were damaging the ozone layer, especially over the Antarctic. The discovery revealed an extreme danger, because the ozone layer functions like sunscreen for the planet. It ensures that a large amount of the UV radiation hitting earth is reflected back out into space. The scientists discovered that contrary to all scientific prognoses, a third of the ozone above the Antarctic had disappeared. And this had taken place within a span of just ten years. The Montreal Protocol was implemented with strict restrictions on the use and trade of CFCs, driven not least by the US, whose industry had a strong financial interest in bringing its alternatives to market. The protocol successfully prevented ozone depletion: Researchers estimate that the ozone layer will have completely regenerated itself in the second half of the twenty-first century. The example of the saved ozone layer shows that climate protection does not have to harm the economy, but that, quite to the contrary, companies' creative solutions can contribute significantly to climate protection.

The first COP, the "Conference of the Parties," which is the conference of the Intergovernmental Panel on Climate Change, took place in Berlin in 1995. Angela Merkel, back then still minister of the environment, played a leading role.

Later, stubborn resistance led to the exit from nuclear energy, of course also suddenly motivated by the nuclear catastrophe at Fukushima. This was not a political success for the climate in everyone's eyes; but it was no doubt an environmental one.

Historically speaking, the decision to finally exit nuclear power marked the beginning of the rise of the solar industry—largely driven by German politics as well as a visionary German energy economy.

Yes, a lot has happened in the past decades. Many people have worked tirelessly to ensure that the situation is not even *more* serious today.

Imagine once again that the challenge of the climate crisis is a marathon. That would mean we finished half a mile in the past thirty years. And that was a hard-fought half-mile, because in between we kept going backward (the many aborted climate negotiations, the coal plants that were connected to the grid against all resistance, the lost battles over conservation areas and deforestation moratoria). We are moving forward—yes. But in the context of the marathon ahead we have not moved much at all.

LUISA The sound boxes that we had planned to borrow hadn't been charged overnight. By the time we found out, we had a half-hour to the beginning of the strike. The negative temperatures in Berlin meant that no one but us had found their way to the meadow in front of the parliament building, where now, on the morning of December 14, 2018, we were standing around, somewhat at a loss, asking ourselves whether the whole thing wasn't just going to be a huge fiasco. A few days before we had shot a video in front of the building and shared it on all our channels. We also sent an endless number of messages through WhatsApp groups. For hours, I and many others had spoken with students in person and over the phone; kids who didn't know how to convince their teachers or parents that all of a sudden they didn't want to go to school on Friday, because there was going to be this "Friday thing." Nobody knew what to expect. There had been no Fridays For Future protest in Germany until then, nor had anyone heard of it or what it was.

The first Friday was therefore more than just a test. And here we were. Without sound. How do you engage people who have come to participate in what was billed as a major strike (if they do come at all) without music or mics? Where do you find a wireless sound system on Friday morning at 9:30 that is big enough to reach hundreds of people? Or are there outlets somewhere in the ground in front of the Bundestag? So many questions I had never had to deal with before until we were trying to breathe life into a movement in Berlin with somewhat uncertain looks and ice-cold breath. A reporter for the *taz* newspaper who was with us that morning wrote in his notebook and mumbled "ten minutes to the strike, four people present, no sound system."

An hour later just over three hundred people stood in front of us, shouting loudly, singing, clapping—the first Fridays For Future strike in Berlin. Somewhere, helpers had unearthed a rusty generator to supply electricity to the sound system. It was booming almost as loudly as the sound system, but that didn't matter. The air vibrated anyway. Because it felt big, this actually quite manageable crowd in front of the German parliament. Not because the cameras were there, but because we were there. Because so many young people had come, without expectations, but with the great willingness to become part of the transition. Because people came who believed that it could not go on like this. Who were united in the conviction that this intolerable climate policy had to end.

I had the utmost respect for every single person who demonstrated with us on that December morning in front of the German parliament against stealing the future and destroying the climate. With cold hands and feet. They all came without knowing what the outcome would be; a success or just a missed lesson and the resulting trouble with parents and teachers. They came because they no longer wanted to be cheated out of their future. These people came because they felt that change was possible, doable. It is so easy to underestimate people.

3 WE LACK A UTOPIA

LUISA "I just have a quick question, real quick." A thirteen-year-old girl comes to the stage and taps me on my arm. We're in Munich, it's summer. I have just explained to the four hundred people in the audience why it is all rather hopeless. Now many people are bustling around, it has gotten late, and I have been last to speak. These moments, when the lights go back on and the pressure falls off, when presenters and listeners mingle and all somehow belong together, these are moving moments. Even if sometimes they are clouded by the exhaustion that arrives when the adrenalin drops off. Breathing quickly, the girl moves a few strands of her hair from her face. Two of her friends are standing at a little distance, looking shyly to the side. "I just wanted to know whether you can even imagine having children?" I have to swallow hard, although I hear the question so often. After events, especially from young girls and women. They often tell me that they have heard of the so-called BirthStrike movement in England. That they feel it is irresponsible to bring children into this world. Even though they are still children themselves.

Children are the future. If you feel that you would rather not have children that means that you cannot form a positive picture of the future. This doesn't only apply to thirteen-year-olds. Everywhere you look: a lack of imagination. We ask ourselves: What is making our parents so unimaginative? Their own youth was marked by the Cold War, a time during which two systems, two ideas of society, stood irreconcilably opposite each other. The contest between capitalism and socialism was the contest of great visions, which promised, one way or another, the realization of a utopia. Here the classless society without exploitation, there the open society in which people exchange their ideas and products on the market.

The competition between the systems motivated both sides. Socialism in the East was spurred on by the colorful world of consumption in the West to offer more to its citizens and eventually to introduce reforms that accelerated its demise. Western societies reacted to the promise of equality in the East with cautious redistribution. The rich should not be too rich, the poor not too poor, and the middle class should be as large and prosperous as possible.

But already in 1984, five years before the fall of the Iron Curtain, the philosopher Jürgen Habermas warned: "As utopian oases dry up, a desert of banality and bewilderment spreads."[1] That's exactly what happened. Following the collapse of real existing socialism thirty years ago, neoliberal capitalism, too, is in a deep crisis. Global warming, environmental destruction, and growing inequality are the consequences of an unleashed economy that is focused on profit and quarterly earnings above all, and not on the well-being of humans and nature. The rise of right-wing populism is another symptom of this crisis.

The world in which we are growing up is marked by an astounding lack of imagination. Where are the inspiring images of the future, and the stories that stand as models at the horizon of a societal transformation? It is true that we know today that we cannot go on like this. But what is to replace the current systems in the long term is more than unclear. What comes after fossil capitalism?[2] Even the goal of a zero-emissions society has a big catch. It provides a form, a geophysical framework so to speak. But the content remains unknown. The question of what means will get us to the end remains part of the negotiation. What makes things even more difficult is that the notion that we are actually on the right path is still surprisingly widespread.

LUISA "If you look at this country's climate policies there is no reason to hope that we are going to comply with the Paris Agreement or indeed that we will achieve anything commensurate with the scale of transformation that this crisis demands." I am listening to myself speak and find myself frowning. I have said this sentence so many times already, over and over, in personal conversations, in talk shows, on stages and panel discussions, at corporate headquarters and in ministries. Each time, I get the impression that I am only stating the obvious. Every time, I look around the room in the hope that someone will convince me otherwise.

Sometimes the glances that I meet are perplexed, sometimes annoyed, clueless, or upset. But with a few exceptions these glances are saying in various nuances: "We are working on it." As though everything would be fine if we just keep going in the same direction.

When I talk to Christian Lindner, today's German finance minister, in his Berlin office, he leans back in his chair, forms a circle with his thumb and forefinger, and declares that more innovation, more openness to technology, and even more "market" would take care of things.

Peter Altmaier, the former minister of economy, still had scrambled egg in his mouth as he gesticulated wildly to me during our *Spiegel* debate, arguing that strong economic growth is the precondition for everything else anyway. And since that was going so well at the moment, as his colleagues from the other economic institutes had confirmed, we simply had to "continue" with it. Only perhaps with a bit more ambition.

Svenja Schulze, our former minister of environment, throws me a sympathetic look during a Monday evening talk show as she leans far across the table and adds with a concerned look, "We still have to do more." Before her, Volkswagen CEO Herbert Diess had explained that VW, too, would "do even more" now. More what? Build even more cars. Electric cars, he said, because they are somewhat less polluting. The guys around the table nod emphatically, because they all agree: Yes, that's how the future can work out. Keep it up—just a little more.

THE END OF HISTORY?

In 1979, French philosopher Jean-François Lyotard declared "the end of grand narratives."[3] Ten years later, US political scientist Francis Fukuyama declared "the end of history." His thesis was that the combination of liberal democracy and market economy had obviously prevailed as the best possible form of society; the great countermodels to it, fascism and communism, both had failed in the twentieth century. Apparently, he wrote, modern liberalism was more successful than its historical alternatives at working on and overcoming social contradictions. After the market principle had become the norm worldwide, the future was simply a question of solving technical problems; the political scientist counted environmental questions and the satisfaction of consumer needs among them.[4]

Fukuyama was not proven right. From today's perspective the end of history can at best be expected as the end of human life as we know it. At least if we don't change our ways. The climate crisis is not a technical problem. Fukuyama should have known about the danger of global warming in 1989 when he wrote his essay, and in 1992 when he turned it into a book.

The Western industrial nations, where the combination of liberal democracy and market principle developed and spread, are the same countries that have contributed the most to the climate crisis.[5] This can be measured in their cumulative CO_2 emissions—that is, by the amount of additional greenhouse gases that have been added to the atmosphere through human activity. The cause of the climate crisis is the sum of anthropogenic greenhouse gases that have accumulated in the atmosphere since 1850.[6] The largest emitters of these gases are the United States, China, the former Soviet Union, and—already in fourth place—Germany.[7] The United States and the countries of today's EU together have emitted more than half of all emissions. Germany alone contributed 3.8 percent—and this even though it currently accounts for just 1.1 percent of the world's population. The Potsdam Institute for Climate Impact Research therefore speaks of a "historic responsibility" of countries like the United States, Germany, but also China (13.5 percent share of temperature rise), India (7.5 percent) and Russia (6 percent).[8] When we speak about responsibility in the time of the climate crisis, it is not only about getting to net zero emissions as fast as possible.[9] Previous generations in this country have contributed more to this disaster than the people in the Global South, who are, however, more heavily affected by the consequences.[10] That is why we also bear greater responsibility to lead us out of this misery.

NO PLANET B

If all people were to indulge in a lifestyle like the average German's we would need three planets.[11] And it seems as though the world is nevertheless on the disastrous path toward that goal: Global greenhouse gas emissions have risen by over 40 percent since 1990.[12]

Today, as then, there is no reason for the optimistic "carry on" attitude that characterizes Fukuyama's thesis of the end of history. "There is no Planet B"—the demo slogan sums it up.

There is no serious justification for the claim that decision makers were unaware of the dangers of rising emissions. The belief that the self-regulating power of the market is an efficient panacea has apparently stunted the imagination to such an extent that members of the German federal government publicly describe political decisions that bow to the logic of the market as having "no alternative."[13]

No matter whether it was about the reaction to the financial crisis, raising the retirement age to sixty-seven, or the military mission in Afghanistan: The federal government adopted this political rhetoric of justification. In 2010, the word "alternativlos" (without alternative) was rightly declared the "Unwort des Jahres" (most inappropriate word of the year);[14] the jury argued that it "inappropriately suggests that there are no alternatives in a decision-making process from the outset and thus no need for discussion and argumentation."[15]

But that's what it actually should be about: the discussion of alternatives and the exchange of arguments. It should be about imagination and maximum creativity in forging novel plans.

"How could it happen that nobody foresaw this crisis?" the queen asked the British Academy after the world financial crisis. The scholars' answer reads like a blueprint for the handling of the climate crisis: It was the result of "a failure of the collective imagination of many smart people, in this country and also internationally, to understand the risk to the system as a whole," they told her.[16] So it was not only the greed and recklessness of individual players at the financial markets that led to the crisis but also the failure to see the consequences for the system as a whole.

"Too big to fail": People rely on large systems without noticing when the floor begins to sink under their feet. The same is true for the nuclear accident in Fukushima. In March 2011, a tsunami triggered by an earthquake led to the melting of a nuclear core at the Fukushima Daiichi nuclear plant. About 18,500 people died as a consequence of the tsunami, and estimates suggest that up to 150,000 people had to evacuate their homes either for the short or long term due to radioactive contamination.[17]

At first glance, these are the victims of an unexpected flood disaster that led to an accident inside a nuclear power plant. At second glance, the philosophers Silja Graupe and Harald Schwaetzer argue, this catastrophe is also the result of "complacent and at the same time deceptive security" of the power plant engineers. Graupe and Schwaetzer write that even at the moment of the earthquake that led to the tsunami, the responsible engineers were firmly convinced that the plant and nuclear power in general were basically safe: "This belief turned into absolute bewilderment when the tidal wave crashed into the power plant: Suddenly, the engineers were sitting in total darkness in a normally brightly lit control room. In the glow of their flashlights, they saw the radiation readings on their instruments rise to incomprehensible heights before these instruments failed altogether and shut down."[18]

The people in charge simply could not imagine what would happen, so they considered themselves to be completely safe until the event occurred. In order to prevent a catastrophe like Fukushima from happening in the first place (for example, by refraining from building such a plant in the immediate vicinity of the sea), the two philosophers state, "it would have required the ability to make conscious decisions specifically for this place and time . . . and to actually imagine the suffering and problems of people and nature not only in and around Fukushima, but for the entire world and the countless generations after us, and to approach any construction on the basis of this imagination."[19]

Thus, it was not a purely technical problem (walls that were too thin or low, safety technology that was deficient, early warning systems that were not fully developed) or the unexpected tidal wave that led to the disaster. The cause was much more a lack of the ability to imagine the dramatic consequences, to accept their possible occurrence—and to act accordingly.

What does this have to do with the climate crisis? The consequences of the climate crisis cannot be understood in terms of singular events such as the Fukushima I accident—they are expressed in a multitude of different occurrences: extreme weather, heat waves, floods, storms, melting glaciers, rising sea level, species extinction, and so on. This also means that the connections between an action and its concrete consequences are rarely directly tangible and cannot be recognized in their causal relationship.

Therefore, the conclusion we have to draw for dealing with the crisis is all the more dramatic; it is a matter of exercising an imagination that can grasp the consequences of our actions for future generations and other regions of the world.

And then you turn on the TV. Armin Laschet[20] with Anne Will[21] on the evening of the European elections in May 2019: Sitting in an antiquated-looking leather chair, the deputy chairman of Germany's most powerful party says that he can't explain either why climate has so "suddenly" landed on the agenda. Perhaps he simply hadn't noticed what had been happening around him in the five months before. Ongoing climate crises had paralyzed the world; 2018 had been declared one of the warmest on record—and the year with the highest total global emissions in human history. Young people had therefore demonstrated in every German state for real action against the climate crisis.

A few months earlier: Transport Minister Andreas Scheuer at the New Year's reception of the German Association of the Automotive Industry. He is pleased to be speaking "among friends" and categorically rejects a debate about speed limits, the health risks posed by particulate matter, or higher fuel prices. The focus should be on the growth of electric cars.[22] At the same time, Economics Minister Peter Altmaier constantly preaches that climate protection should not be pursued "at the expense of prosperity and jobs."[23]

Of course, electromobility is an important building block on the road to a sustainable society. But to overcome our disastrous dependence on fossil fuels we need to rethink urban mobility instead of flooding the country with new electric cars and leaving everything else as it is. Free local public transport, as in Luxembourg, or a (virtually) car-free inner city, as planned in the city of Bremen, are first steps in this direction. And these investments in local and long-distance public transportation create new jobs.

The Laschets, the Scheuers, the Altmaiers of this country—they are exemplary for a political leadership and decision makers who spread innovative spirit and euphoria for the future in such homeopathic doses as if we still had an eternity. Instead they keep emphasizing what all might go wrong if the AfD [Alternative für Deutschland (Alternative for Germany), a right-wing populist party], the yellow vests, low-income earners or top taxpayers, property owners, tenants, or industry suddenly went to the barricades; and why it is so important to act with moderation.

LACK OF IMAGINATION

"We have forgotten how to dream,"[24] is how social psychologist Harald Welzer sums up the situation. This does something to a society, it does something to a generation: If one is not taught to dream, to dream big, to think big, to develop visions, and to break free from the status quo, where will the urge to roll up our sleeves come from? Do we have to draw everything from our own resumes?

It is not an option to leave the future to right-wing demagogues who, in their ethnic and racist fanaticism, wish for the return of a world that never existed. But neither is it an option for a technocratic elite that has made itself comfortable in the fortress of "no alternative" policy proposals. In this context, the journalist Robert Misik speaks of a "myth gap" among progressives, a myth gap that needs to be overcome. He quotes from a discussion on the platform opendemocracy.net: "While our instincts are to counter the lies and distortions with facts and data, the real challenge is to beat the right in the field of myth production and storytelling." Misik concludes, "What is needed, then, is a narrative into which all the facts, data, and positions on factual issues can fit."[25]

Myth, narrative, vision, utopia. Whatever we call it, the world of tomorrow, the world we dream of, should be a place of longing. If we lack this place, we run the risk of losing sight of the goal in the tough and complicated everyday life of democratic processes. We cannot pull this vision out of a hat either. If it were that simple, we would have solved the problem already.

4 THE CLIMATE CRISIS IS NOT AN INDIVIDUAL CRISIS

- How did humans react to the climate crisis?
- They succeeded thanks to collective enlightenment and the greatest spirit of their leaders to revolutionize their economic system.
- Really?
- No, just kidding. They banned plastic straws and ate organic food once in a while.
- And?
- That's it. They went extinct.
  (A Twitter user)

LUISA, on a panel discussion in Hamburg: I am exhausted. I'm sitting on a sagging sofa answering questions about the future of the climate movement. I had tried to get some sleep on the train ride to the event, but I was recognized by the person sitting next to me. For him, this was the opportunity to explain to me for an hour and a half why we should be worried about Chemtrails. I had tried to signal to him—politely but in vain—that I was busy.

Now it's dark in the room; so dark that you almost fall asleep before anyone says a word. Two and a half weeks before the European elections, the schedule has become increasingly full and the nights shorter and shorter. There is no coffee. Fortunately, a scientist is with me on the panel. She can talk about the climate catastrophe in a way that you rarely hear. I am deeply impressed by this mixture of wealth of knowledge, forcefulness, and concern—and also grateful that it is a woman who is talking like this. We will be hearing more from her—Friederike Otto.

As in almost every interview and on the vast majority of panels and talk shows, there is of course a question shortly before the end that has long since become a must in every climate discussion: "And what do you personally do for the climate in your everyday life, what do you do without—what can each person do individually?" Before I'm handed the mic, the three others on the panel have already said how much they recommend biking to work (with helmets, of course) and taking a meat-free day a week, and that, of course, you have to go vote, too. But I sit there, squinting into the audience and wondering if I'm the only one who finds this situation absurd.

For almost two hours we had discussed the biggest and probably most complex crisis in human history. We had emphasized how important it is to turn the big screws, to ask systemic questions, to initiate structural change because we have so little time left to decarbonize the whole place. But people are sent home from discussions like this with the completely predictable answer to climate protection in everyday life. That's supposed to be the last thought that's being articulated? That's what's supposed to stick? Ride more bikes and fry more tofu so that we can feel good? This reduces a complex problem, to which we must respond very quickly with a fundamental change in our economy and lifestyle, to a question about individual consumption.

I take the mic and say that I refuse to answer this question in this way. I am generally critical of questions that don't differentiate between collective and individual responsibilities. But it is only on this podium that I realize how disastrous this pattern of conversation is. The fact alone that hardly any interview gets by without this question speaks volumes. Volumes about how overwhelmed we are. Since that evening, I have never answered this question unreservedly. My answer is, "Yes, living ecologically can be great, enriching, and fun. I encourage everyone who has the opportunity to try it. But it's crucial to build pressure together to change the structures."

In the context of the theory of political economy, the climate crisis can be called a "tragedy of the commons." The term was coined in 1833 by the British economist William Forster Lloyd. It was meant to describe a problem that arises when an openly accessible resource is shared by different actors: a well or a common pasture where all villagers are allowed to graze their animals. If out of self-interest everyone uses more of this resource than he or she should, i.e., allows more animals to graze on the pasture than grass can regrow, this results in overuse. This in turn harms everyone, and the community of all users faces a problem. Everyone would like to have more of the common good. But if everyone takes too much, in the worst case there will be nothing left for anyone. In 1968, the ecologist Garrett Hardin wrote an essay that made the expression "tragedy of the commons" famous.

Using examples such as the overuse of national parks, environmental pollution, and the rising world population, Hardin showed why there can be no infinite growth in a world with finite resources.[1]

In the case of the climate crisis, the shared good, "the commons," is the sum of global CO_2 sinks.[2] "Sinks" are the places that absorb and store CO_2. These are primarily oceans, forests, soils, and lastly, of course, the atmosphere. These sinks store the CO_2 that is added to the carbon cycle by humans—primarily by burning the fossil fuels coal, oil, and gas.[3]

The 1,331 tons of CO_2 added to the carbon cycle every second by human activities have to go somewhere. However, the level of global emissions exceeds the capacity to absorb the additional inputs without major catastrophes.

Therefore, they are causing enormous damage: First by the unprecedented load of CO_2 in the atmosphere, which increases the greenhouse effect dramatically and leads to global warming. But in the oceans as well, where CO_2 reacts together with water to form carbonic acid. This lowers the pH value of the water, thus leading to acidification of the oceans and polluting living organisms and plants. The calcium carbonate that forms the shells of mussels and snails, for example, is dissolved as a consequence. However, the excessive emission of CO_2 also leads to a significant decrease in oceans' capacity to absorb CO_2. Accordingly, more CO_2 is released into the atmosphere, which further accelerates global warming.

In short, human-made processes are releasing the unimaginable amount of over 40 gigatons of CO_2 per year worldwide. This climate-destroying gas is overwhelming major sinks, leading to a multitude of small and large catastrophes in a complex dynamic. If less CO_2 were emitted by everyone, the sinks could buffer the human-made input. But because the opposite is the case, the whole system is on the verge of collapse.

The disastrous tendency to overexploit the commons, however, is not a law of nature. The economist Elinor Ostrom has shown how humanity has succeeded time and again in using the common good in such a way that it has been permanently preserved for all. She has collected hundreds of examples where resources have been used sustainably on a regional level, so that common rules of cooperation have ensured the future of the commons to this day.[4]

Ideally, all those contributing to the overuse of the system would join together and decide what needs to be done so that they can all continue to use the shared resource without causing harm. This has already been attempted: Appropriate measures have been agreed to in the Kyoto Protocol and later in the Paris Agreement—only so far without putting an end to the crisis.

The crucial point is that not all parties contribute equally to this tragedy, and that they do not benefit or suffer to the same extent from their respective actions. The hundred companies that are responsible for 71 percent of the greenhouse gases of the past thirty years are sending their animals—to stay with the metaphor—in huge numbers to pasture, enjoy the additional income, and are initially not affected by the long-term damage they are causing. The managers of these companies, the already mentioned fossil elite, could give themselves new rules for their work. They could prescribe to their companies to dig up less and less fossil raw materials—to reduce the number of animals on the pasture bit by bit. But why would they?

Neither the leading companies of the fossil system, nor the people living in the CO_2-heavy states are currently feeling this overloading of the sinks. Others have plenty of reasons to advocate changes that will stop the overuse of sinks, but do not have the power to initiate them. They themselves have few animals on the pasture and therefore cannot have a say.

What could be the incentive for decision makers to break up the structures that are responsible for the disastrous emission of 37 gigatons per year? What appeal would need to be made, and to whom, to transform our CO_2-based economic system? We would have to design energy infrastructures, agriculture, transport, and global transportation systems in such a way that they immediately begin to significantly reduce the amount of global emissions.

Some have understood this and are leading by example: Large companies, such as Bosch, are committed to CO_2 neutrality.[5] Countries like Costa Rica,[6] whose power generation is almost entirely based on renewable energy sources. Political representatives who are resolute in the fight against the climate crisis, such as Helen Clark, the former prime minister of New Zealand, who planned to make the country climate neutral by 2025. But that is not enough. We have to acknowledge that all these initiatives have not made enough of a difference. Not even those initiated by people who are willing to change their lifestyle.

LUISA The year after I graduated from high school, I decided to live in the countryside for a while. After spending almost my entire life in Hamburg, I thought it was time to at least begin to learn how farming worked. After a brief stay at a small sheep farm where I felt a little lost, I eventually moved to a community in Southern England, joining twenty volunteers from around the world who were experimenting with ways of living as a community.

I spent just under two months there and probably learned more than I learned in my last two years of school combined. We had resolved to live as consciously in harmony with the environment and its natural resources as possible. No more than once a week did we drive to a supermarket to buy what we could not grow ourselves. The rooms were divided up functionally. We repaired broken things, we forged knives, brewed alcohol from elderflowers, and composted using a system of twenty different containers. In this way we had developed our own micro-energy system and managed with what we had at hand. The only regular consumption was the beer we drank on Friday nights while playing darts in the pub down the street.

We had given a lot of thought to how this could work: to live a life that maintained our livelihood. It wasn't easy, quite the opposite. Many times, I found it annoying and thankless to spend much longer on recycling potato skins than eating the potatoes themselves. Or spending hours scouring the assortment of small stores to avoid even the last plastic package. It takes a lot of energy to reject the chaotic pace of a world in which the aim is to get more and more of everything. To dare to ask the big question of meaning, even if only on a small scale. On the last evening in my commune, I asked myself if this was it, the good life?

I still had my rubber boots in my hand when I stood on London's Oxford Street a day later. And I will never forget it, that moment when the plastic-wrapped madness of that postmodern reality overtook me. Those crowds of people rushing past me, right and left, close together, sweating frantically, between droning buses and honking cars. Into the stores, heavily laden, on and on—*Summer Special*, *Super Sale*. And everything wrapped in plastic, fully loaded shelves all the way up to the ceiling.

I remembered the many times we recycled the last of the food, sewed and swapped clothes—and had the nice feeling of making a difference. Because, after all, every avoidance of plastic waste and unnecessary consumption makes a difference, doesn't it? As I stood there, in the midst of this hyper-ventilating crowd, it was as if the world was trying to tell me:

"Nice try, honey." At first, I thought the world had forgotten itself; then I realized that I had forgotten the world.

THE LUXURY OF RIDING A BICYCLE

We don't need a panic-stricken renunciation debate, but a debate about the good ecological life. Where the positive climate balance is associated with pleasure and luxury. Moderate consumption that is in harmony with ecological cycles and respect for human rights standards can make a critical contribution to a better future. The best example is the bicycle. Once it has been produced (although the energy balance of production can still be improved), it is designed to last a long time. It has the best energy balance in use and is beneficial to health. And it has a much-underestimated effect of self-empowerment: Riding a bicycle gives you true freedom of movement. The only thing a bicycle needs—apart from a little chain grease—is the time you take to ride it. And owning more than one bicycle does not make you faster.

The luxury of a good and functioning bicycle is representative of a lifestyle whose core is not blind consumption at the expense of the planet (and often one's own health)—but a life that consciously makes use of the most valuable resource of all: time.

All that is needed is the right infrastructure. Yes, it is up to each individual to buy a fast bicycle or not, to ride one or not. However, it is beyond the individual's power to create a bicycle infrastructure so it can be used safely, for example by children. Furthermore, it is not within the power of the individual to create places where bicycles can be parked without being stolen. Or to design transport networks in such a way that bicycles can come close to competing with other means of transport. No one will ride a bicycle to work every morning, no matter how nice, if the ride seems like a life-threatening odyssey through a polluted city.

In a world fueled by fossil capitalism, one's freedom to behave in a climate-friendly manner or not all too often appears to be an illusion, at least as long as one's efforts have hardly any measurable effect on the overall climate balance, and as long as it is unclear what ecological price is paid by each individual product and how ecological costs are priced into consumption.

GREEN GUILT

Green guilt is the term scientists use to describe the feeling that arises when the attempt to "do everything right" chronically fails. When one's efforts to be part of the answer have the debilitating aftertaste of futility. When, after years of diligently separating garbage, you realize that everything gets dumped together again in the disposal systems; when you realize that "fair" doesn't always mean fair; when you start understanding that ecological costs are hidden everywhere.

Besides, the green guilt leads to an absurd competition: Who can make it in this madness in the most environmentally friendly and climate-friendly way possible? Everyone knows that you can never do everything right.

Green guilt is a slap in the face of a society that wants to engage in dialogue, but is not allowed to do so by the real decision makers. And who then, instead of angrily kicking down the door, organizes a zero-waste picnic in the park. A picnic that, in itself, is wonderful and can be fun. But it still doesn't answer the question of how we can transform our world as quickly as we need to.

Meanwhile, the industry carries on quite undisturbed, using the increased environmental awareness for new outlets for products that look handmade or are decorated with chia seeds. "Starting with oneself" is now misunderstood as the *ultima ratio* of individual freedom of action, as if everyone can contribute something to solving the crisis, if there is only the will. In this way, a crisis that is a crisis of society as a whole, or even a global crisis, is shifted into the private sphere.

If, at the end of every discussion about the climate, the question "What can each individual do?" is asked, it shows that our society no longer sees itself as a society, but as a collection of individuals whose political influence is limited to their shopping list. The privatized and individualized climate crisis thus appears as a problem that was triggered by individuals and consequently can be remedied by individuals.

SHIFTING BASELINES

And now? You might think that there's nothing you can do about it, on the whole. That is understandable. So, you withdraw, away from this overwhelming crisis, you focus on your own life and accept your own powerlessness. On a large scale, this creates a powerless society. Literally. The fact that not everyone sees it that way, however, makes things exciting. There is already a critical mass that says, "I want to do something," a mass of people who want to be part of the answer and join the team of go-getters. More and more people consider this prospect worth striving for. And rightly so.

Until we get to the point where a lifestyle with a minimal ecological footprint can be integrated into the everyday life of a normal citizen, a lot still has to change. A whole lot. Because in the end, the climate-friendly consumption decision must always also be the intuitive, inexpensive, and convenient one. In his book *Ökoroutine*,[7] environmental scientist Michael Kopatz describes this state of affairs as a reversal of the prevailing conditions: Today, behaving, consuming, and moving ecologically usually involves extra effort; it is a question of money to be able to afford environmentally friendly alternatives at all, and a privilege to live accordingly.

In the end, the so-called baseline, i.e., the broad social and economic framework, must be an ecological one. Even then, it would still be possible for every individual to behave in an environmentally harmful way; but this would become the exception. And the person concerned would then have to bear the additional social costs. Consider smoking: Those who want to smoke can still do so, but smoking is neither cheaper nor more convenient than not smoking.

The fact that lower-income households leave a small environmental footprint on average can be attributed less to conscious ecological consumption than to enforced frugality. This is not the solution we want. Absurdly, however, the very households that claim to be environmentally conscious are also the ones with the higher carbon footprint.

So as long as the framework conditions are not changed fundamentally, consumer criticism alone is ineffective, because it does not promise a sustainable answer to the big questions. If the transition is to be successful, consumer criticism can be no more than a beginning.

Imagine that only products would be allowed to be made for which the manufacturer can guarantee a closed resource cycle. It should be self-evident that a market should not be allowed to be flooded with goods for which there is no plan for the end of their life cycle, broken or no longer in use. This problem arises in an increasingly digital world, for example, with almost all electronic devices. It is commendable to take your cell phone to the recycling center. Nevertheless, the big adjustments have to be made elsewhere. And this requires a mass of people who organize themselves. By boycotting certain products. Or, even better, by exerting pressure on the streets, in institutions, and at the ballot box. It's simply not enough to buy vegan cheese once in a while.

> "The issue of global warming is, in my view, one of the most important environmental issues that we need to address. And that's why I will try with all my might to achieve as many successes as possible in Berlin, even if the process is often too slow for me, too."
>
> "Let's end with the most important question, Mrs. Merkel: What do you do privately? Do you separate your household waste, do you use the bus and train privately, do you turn off the water when you brush your teeth?"

Mrs. Merkel did separate her waste, at least at the time of this 1995 television interview, when she was still minister for the environment.[8]

5 THE CLIMATE CRISIS IS A CRISIS OF RESPONSIBILITY

LUISA Since my teenage years I have been concerned with the climate crisis. In all these years I have only once considered giving up. To just let the climate crisis be a crisis and turn to other things. That moment arrived one night, three months before this book was first published. I was sitting at my desk in my shared apartment in Göttingen and trying, as I often do, to understand why we hadn't done something long ago. My most productive hours were usually the ones after 12 o'clock at night, when my cell phone went silent and my head quieted down a bit to let my thoughts wander.

The windows were wide open on nights like this, and sometimes there was a slight breeze. My apartment is in the middle of the old town, and the sounds of Göttingen's nightlife echo right into my room. It's the only thing I can hear besides the clicking on the keyboard and the occasional rustling of my notepads. Piled up on my writing table were books by smart people who had written about so much of what occupied me during those months. Decades ago, these people had already warned of the climate catastrophe. What had they overlooked?

A few hours before, I had joked with a friend about whether we would still drink so much maté once we were parents. Whether we could imagine moving back to the areas where we had grown up. We thought about growing old in Germany and whether we would live to see the day when we could tell our children about the old days when people still moved around in cars. The old days, before electric shuttles, air cabs, or full-body beamers had solved these problems for us.

Now it was getting late. I read about Ken Caldeira, the climate scientist who, at the beginning of each semester, asked his students what was the biggest breakthrough in climate physics since the 1980s. It was a trick question—because the answer is: There were none.[1] Ever since scientists like James Hansen had realized what was happening to the planet, the data had become more and more precise, but fundamentally, shockingly little had changed.

I read about the many moments in history when it almost seemed as if states would act in time. And of the scientists who, long before I was born, testified before the United States Congress about climate change. I read about the first UN climate conference in Berlin in 1995, a year before I was born. But none of the many conferences had been able to stop the rise of CO_2 in the atmosphere. It seemed that there was one lesson above all others from the past thirty years—that the climate crisis was too big and too complex for humanity. That perhaps this was the fight we were going to lose.

I consider myself a fairly positive person. I'm fun loving, I like challenges, and I'm convinced that people fundamentally mean no harm. And I do believe that everything will be all right in the end. But the knowledge of so many futile efforts in the past, the knowledge of the state of the world, and the lack of political will to change it, brought me to the limits of my confidence that night. The chances that we would still be able to do enough to avert the catastrophe were simply too small. Vanishingly small.

Now, in a moment that could not have been more peaceful, this injustice overtook me. It came with a force that took my breath away. It was half past one in the morning, and I heard a glass bottle break on the street outside.

I remembered how shortly before I had talked quite lightheartedly about what my life would be like in the future. That must have been a moment of gullibility; a moment when I had forgotten that this future would never exist. Because everything was up for grabs: to grow up liberated and mature, to gain experience, to become a mother or even a grandmother.

For the powerful of this world, my life was a mere burden in their everyday business, an inconvenient open bill that no one wanted to pay. The unwelcome guest, the detail that was best ignored. A responsibility that no one wanted to live up to. Everyone had more important things to do.

We would not be in this crisis today if the question of responsibility were not so complex, and we will not even begin to overcome this crisis until this question is answered.

The notion that we bear responsibility for future generations was not our idea. We got it from our Constitution. Since October 27, 1994, it has said there that, "Mindful also of its responsibility toward future generations, the state shall protect the natural foundations of life and animals by legislation and, in accordance with law and justice, by executive and judicial action, all within the framework of the constitutional order"[2] Article 20a of the German Constitution. Thus, for twenty-five years, nature conservation has not only been a moral obligation for politicians, but also an official state goal. For the record, this national objective is not just about environmental protection and animal welfare; it is also about protecting the natural foundations of life "in responsibility for future generations."[3]

Many Eastern European constitutions, as well as the constitutions of the five new German states, speak of the rights of people who have not yet been born. The Constitution of Poland of April 2, 1997, states: "The state shall pursue a policy that ensures the ecological security of present and future generations."[4] Already the preamble to the 1993 constitution of Thuringia speaks of "responsibility for future generations."[5] The constitution of Saxony, adopted a year earlier, formulates the norm: "The protection of the environment as the basis of life is, also in responsibility for future generations, the duty of the state and the obligation of all in the state."[6]

Why do we insist so much on the responsibility of politics? It is not mere idealism when we young people accuse the powerful of not living up to their responsibility toward the environment and future generations. We can refer to our Constitution, the Basic Law. It is the politicians who must now act urgently.

When we talk about responsibility, we are talking first and foremost about an entire generation's failure to act. In the frenzy of joy following the end of the East-West conflict, they forgot about the future of their children and their children's children.

Why could all this be ignored for so long? The environment and the future generations not yet born have no voice in our political system. They cannot vote and they cannot protest.

Article 20a is not about welfare, but about the basis of human life on this earth. With its introduction, the so-called *Vorsorgeprinzip* (precautionary principle) was anchored in the German Constitution. This principle basically says: Don't break anything if others have to pay for it afterward. And if you are not sure how great the damage of an action will be, then it is better to avoid it. The philosopher Hans Jonas formulates it somewhat more elegantly in his book *Das Prinzip Verantwortung* (The Principle of

Responsibility). He calls it "the rule that the prophecy of doom is to be given a greater hearing than the prophecy of salvation."[7]

When it comes to the precautionary principle, the climate policy track record of the Federal Republic of Germany has been disastrous so far. People who live in other parts of the world or belong to future generations have to pay for what we are destroying every day with our mobility, energy, and consumption habits. That's why we have to change things now, even if we do not feel the consequences ourselves, or only much later.

Hans Jonas calls such action "responsibility for the future." We owe much to Jonas when we talk about the precautionary principle, the rights of future generations, and those of the environment. As early as 1979, when *The Principle of Responsibility* first appeared, he warned about the technical capacity to permanently alter the conditions of nature and future generations. The "promise of modern technology," Jonas wrote, had "turned into a threat" because the excessive success of science and technology now enabled us to destroy the physical foundations of humanity.[8]

In a world where our actions have an impact on people on the other side of the globe or at the end of the century, responsible action must look beyond our own temporal environment. In this, we all bear responsibility, but not equally so.

Because, according to Jonas, political decision makers gain power precisely through their decision to assume responsibility (for example, by standing for election and taking office). The kind of responsibility that comes with power, therefore, does not arise from what has been accomplished, but from what "must be done." Thus, the office of politicians entails a "duty of power," which obligates them to protect the area of responsibility entrusted to them for the common good. Those who stand for election must ensure that in the future, too, people will be able to live here and go vote.[9]

Jonas compares political responsibility to that of parents for their children. According to Jonas, the child is the "original object of responsibility."[10] Parental care for the child expresses the natural responsibility that has arisen irrevocably from the role of parents as parents. This form of responsibility is irrevocable, global, and permanent. It cannot be discarded.

And just as the parents are responsible for the well-being and survival of the child, the chosen responsibility of the politicians is to provide for the well-being and survival of society. Unlike parental responsibility, political responsibility can be relinquished at the end of a term of office, but as long as Merkel & Co. sit in parliament or even run the government, their actions are to be measured by how successfully they fulfill this task.

But what do we do when politicians are completely unwilling to fulfill their duties? How can we defend ourselves against the irresponsibility of those responsible? When people do not fulfill their self-imposed responsibility that creates a vacuum, a space of irresponsibility, a space of negligence. We call this a "responsibility crisis." Because when those who bear responsibility do not live up to it and no one else does, a void opens up that gives free rein to those who recklessly damage natural spaces, livelihoods, and social peace. We will not solve the climate crisis if the distribution of responsibility is not reformed, or even revolutionized.

At this point, the eyes are on all those who are aware of this crisis of responsibility: It is up to them, to all of us, to contribute to changing the initial situation in such a way that the transfer of responsibility can be resolved within the meaning of Article 20a. We see two ways in which the responsibility crisis can be addressed.

DEMANDING RESPONSIBILITY FOR THE FUTURE

One option is to demand that the state assume responsibility. This can be done by suing, but that is not necessarily easy.[11]

Legal responsibility is not about something to be done, but about the causal attribution of responsibility for acts committed. The damage caused is to be repaired as far as possible. The question then arises: Who is guilty?

With regard to the climate crisis, the question of assigning blame initially seems hardly applicable, since the causal link between concrete actions or nonactions and their climatic consequences is difficult to prove. The complexity of interrelationships, of causes and effects of global warming, make the assignment of responsibility enormously difficult. As already mentioned, we all contribute to the aggravation of the climate crisis with our consumption behavior and our means of transport—but it will be difficult to establish that it was precisely my flight to New Zealand that triggered the heat wave in Central Europe.

Nevertheless, there are numerous attempts to assert the legal responsibility of governments, companies, and individuals in court. One example from Germany is the lawsuit that lawyer Roda Verheyen, together with Greenpeace, led with three families against the German government. The families sued the government for failing to meet Germany's 2020 climate change targets, seeking legal action to compel the government to step up efforts to meet the targets and help protect the threatened livelihoods of farming families. In their complaint, the plaintiffs invoked Article 20a of the Constitution.[12] Later on, the same lawyer filed another case with young climate activists. As noted above, in its historic ruling of April 29, 2021, the German Constitutional Court ruled in favor of the plaintiffs. A similar lawsuit in October 2018 required the state government in the Netherlands to do more to protect the climate. In the second instance, the court found that the Dutch government had violated its duty of care.[13]

In Switzerland, a group of women over sixty-five who are feistily calling themselves "Die KlimaSeniorinnen" (female climate elders) is suing the government because the reduction targets are too low to meet the Paris climate goals.[14] In Pakistan, in turn, seven-year-old Ali is suing his government after it approved new coal mining areas.[15] In the United States, twenty-one young people sued their government because its anticlimate policies violate their rights to life, liberty, and prosperity.[16] The European Union is also being sued by ten families from Portugal, France, Italy, Greece, Germany, Romania, Fiji, and Kenya, seeking a court ruling that the 2030 climate targets are too low to guarantee fundamental rights and to comply with international agreements such as the UN Framework Convention on Climate Change.[17]

If these lawsuits are successful, the responsibility for the future we are calling for with Jonas could increasingly become a legal responsibility. This would presumably also entail a fundamental change in the setting of political priorities: From now on, decision makers and those responsible would no longer be able to dismiss, sit out, or downplay as "youthful idealism" the uncomfortable questions surrounding the rights of young people and future generations. They would have to face up to these questions or risk having to answer for them in court.

Few people and interest groups, young or old, have the ability to sue. It is expensive, complicated, and time consuming to assert one's rights in court. But political responsibility can also be claimed beyond the courtroom—for example, through protests and creative ways of attracting public attention.

THE PARABLE OF MOURNING THE FUTURE

"After Noah had returned home from his hundredth walk of warning, he could no longer conceal from himself that to continue on, never advised by his God, a hundred times, and each time on his own, no longer made any sense. For yet again, he had not succeeded in recruiting even a single one of his fellow citizens to build his arks. Yet again, the few on whom he had been able to impose himself had been greedy for nothing but the very latest; and they, too, had hid away when he had come to them with the flood warning (with 'his flood,' as they called it), because they had already heard about it yesterday and the day before, and the day before that."[18]

With these somewhat antiquated-sounding words the philosopher Günther Anders begins his parable about "mourning the future."

Noah, who wants to persuade his fellow citizens to build several arks in view of the impending flood, initially despairs of their ignorance. Since they are only "greedy for the very latest," they do not heed his flood warnings. What Noah does next constitutes a blatant break with the conventions of his time: He dresses himself in sackcloth and ashes, in other words, in a garment of mourning, and publicly weeps for the dead of tomorrow, i.e., those who will have perished in the flood. (Yes, exactly—future perfect: He weeps for the future's past.) Driven by curiosity, the people finally listen to him. They are outraged and paralyzed when hearing him sing the song of the dead. At last, Noah's message gets through to them. He succeeds in persuading his fellow men to build the ark. Thus, he saves creation. His stirring spectacle saves life on earth. End of story.

Günther Anders wrote his parable in 1961, in the face of the threat of nuclear war. But it seems more relevant than ever today. The climate crisis is here, and in some places the situation is marked not only symbolically but quite literally by the danger of which Noah warned his fellow men. Floods and rising sea levels threaten not "only" coastal states like Bangladesh and island states like the Marshall Islands, but also, for example, the Netherlands, northern Germany, and the regions around the major rivers in the United States. Some of them are in danger of sinking into the sea if the globally rising average temperature is not stabilized.[19]

How might Noah get through to his fellow human beings today? Who are the Noahs of the twenty-first century who will use unconventional methods to persuade others to act? Is Greta Thunberg such a figure?

During her speech, delivered in December 2018 at the United Nations Climate Summit in Katowice, she, like Noah in the parable, lamented a major failure: Over the past twenty-five years, she said, people have failed to get their governments to act against the threat of catastrophe. Greta, too, has given up pleading. Instead, she points to the failure of political leaders and accuses them of irresponsibility: "We didn't come here to ask world leaders to worry about our future. They have ignored us in the past, and they will do it again. We came here to let them know that change will come—whether they like it or not. People will rise to the challenge. And because our politicians are behaving like children, we have to take on the responsibility they should have taken on long ago."[20]

When Greta began sitting in front of the Swedish parliament in August 2018 instead of going to school, she defied perhaps the only expectation a society has of its students. The school strike shook people up like Noah's death song. Even though Greta only explains what has been widely researched scientifically for forty years. What she started and what we have taken up together with millions of pupils and students worldwide is perhaps the necessary wake-up call that will move people to build the arks.

Noah was very old when he fought for the lives of his descendants. This is the difference to the parable: Today it is the youth themselves who turn against the decision makers. As young people, we lament the suffering that we and our descendants will experience in the future. Therefore, we no longer need Noah to turn back time and mourn the future in order to make people aware of the impending apocalypse.

That's what's happening at Fridays For Future, and nothing else is meant by #actnow: Hey people, you volunteered to run for office. You put yourselves up for election. Now you have a responsibility for our future. Become aware of this responsibility, and act already!

INSTITUTIONALIZING RESPONSIBILITY FOR THE FUTURE

The second way of tackling the responsibility crisis: If the responsibility for future generations is so great, but so few people are willing to take it on, an institutional framework must be created to spread this responsibility over more shoulders. What would happen if bodies were introduced to reliably implement this responsibility? For example, in the form of an advisory council for federal, state, and local governments that would champion the interests of the environment and future generations? Veto rights or other instruments could be used to ensure that such a body is not simply ignored. Such an advisory board could also serve as a body that reviews legislation for its "future fitness," i.e., its implications for future generations—and, if necessary, introduces its own legislative proposals.

Since 2011, the Hungarian parliament has elected an ombudsperson for the rights of future generations. The ombudsperson is also the deputy of the commissioner for fundamental rights, who in turn is responsible for the rights of children, national minorities, vulnerable groups, and the interests of future generations. Such representatives for the rights of future generations take the categorical imperative of our present into account. And they do so in the way that philosopher Hans Jonas formulated it, following Immanuel Kant's categorical imperative: "Act in such a way that the consequences of your action are compatible with the permanence of genuine human life on earth."[21]

6 THE CLIMATE CRISIS IS A CRISIS OF COMMUNICATION

LUISA The hall went dark, the buzz of voices slowly died away. Two thousand people were looking at the stage of a big concert hall called *Friedrichstadt-palast*,[1] cameras could be heard humming, and some took out their smart-phones. A technician adjusted my microphone. My preparation time could be called "athletic"; it ended up taking as long as the bicycle ride over here. A week earlier, we had gone on strike with almost two million people world-wide, and immediately afterward we had occupied the European Parliament for twenty-four hours. Shortly afterward, the Greens had won a landslide victory in the European elections. There was a lot going on, and I hadn't had the time to write a speech. I was shifting uneasily from one leg to the other.

The spotlight came on, and the event began: "I am pleased to welcome Luisa Neubauer, one of the young climate activists from Fridays For Future. These young people are loud, they don't want to wait, they are saving the climate for us. Let's give them a warm welcome." Even as I was walking onstage, I was wondering when we would finally start to find a language that matched the climate reality and move beyond the climate savior narrative or at least breathe life into it. I took a deep breath. "Hi, I'm Luisa. Thanks for having me." We still have so much work to do. It all begins with words.

It sounds cute, "the climate saviors": the people with the big heart, the pro-tectors. The ones who put themselves in front of the planet and protect it from evil forces; from which ones exactly, we don't really know anymore. In case of doubt probably from Donald Trump. But that doesn't really matter. A little pat on the back can't hurt. Animal rights activists protect animals, refugee aid workers help refugees, and climate saviors save the climate. We are the ones who fight against rising climate graphs, save the atmosphere and the polar bears—and, as some seem to think, the environment in general. But such ideas do not get to the heart of the crisis. Because it's really about people and how they deal with their livelihoods.

THIS IS YOUR CRISIS, TOO

Yes, we are experiencing a crisis of the climate, which is leaving behind its familiar tracks, accelerating from one extreme to the next and in the process making people forget everything they have ever experienced. But in essence, this is no longer a crisis of the climate—it is a crisis of humanity. It is no longer "just" a question of whether ecosystems will remain intact in the long run while they are subjected to the stresses and strains of rampant fossil capitalism, which is disrupting nature and climate systems with ever-greater aggressiveness. It is also no longer "just" a question of which, in politicians' language, "system-relevant" species can still be protected in time from final extinction, in order to ensure the long-term nutrition of humans (as in the case of krill, the crustaceans in the sea, which are considered the main staple of many marine species, or of bees on land). It is no longer "just" a matter of sea levels rising by 3.3 millimeters[2] per year, which will endanger more than 800 million people living along the coasts in 2050.[3]

We have already reached a point where the question is whether human life will be possible on planet earth in the long run at all.[4] The planet will continue to exist, albeit under greatly altered dynamics; it does not require our protection to exist. What needs to be protected is humankind, from itself, from the blind self-destruction rooted in our resource-intensive way of living.

But what if these messages do not reach people? What if these messages are ignored, relativized, and swept under the carpet? That is how a society can close its eyes for decades to the inevitable consequences of sawing off the branch it is sitting on.

LUISA I first heard about the emergency slaughters in the summer of 2018. Those were the days when the whole country was talking about the dried-up river Rhine, about crop failures and the devastating forest fires in Greece. There was also discussion about farmers demanding compensation. They did not talk about the details. For example, about the fact that tens of thousands of calves and cattle were slaughtered on these hot days because the farmers could not find any more feed for them.

The fact that factory farming had become so dominant in Germany thanks to government incentives was already a scandal in my eyes. But the idea that animals were now being "disposed of" en masse, because the system had reached its limits, was almost unbearable. I wanted more people to know about it, so I contacted an affected farmer to report on the slaughters.

When we finally found a moment and the tractor stopped, the family man, whom I called Dirk in my report, told me about his life. About his farm in northern Hesse, about his wife and two children. He told me how he drove out three times a year with the tractor to harvest hay, and after each harvest, he would press four hundred and fifty bales of hay.

That summer, which was later declared the second "summer of the century" in just fifteen years, he drove to the pastures only once. He talked about what it was like to stand in the fields that his grandfather had farmed, that had once been lush green pastures, but were now brownish and withered. This year he harvested just enough hay for two and a half bales. That won't feed a calf. He told me how he had to fire people he had worked with for thirty years.

The local farmers used to meet every Thursday evening in the pub for two rounds of beer and corn schnapps, and talk about how they were getting by. Since that summer, the regulars' table had been more sparsely attended each month. The farmers were forced to give up, one after the other. The big agribusinesses were already waiting at the farm gates, picking off family farm after family farm. People who also have families to feed.

But to Dirk the problem did not seem to be a political one. His problem was this summer, which joined a series of exceptional seasons. They simply could no longer rely on the fact that summer was simply summer. It used to be different, he said, at least, as long as he remembered. His son would probably not be able to take over the farm. It just didn't add up, whichever way they looked at it.

For two hours I listened to this father, husband, and farmer who knew that one day he would have to sell his life's work. Not because he hadn't done well, or because there was no longer a demand for his crops, but because the weather had changed.

I sighed, muttering something like, "It's so sick, this climate change." Dirk agreed with me energetically, "Yes, the climate change thing, that's bad too, you hear more and more about that." I replied that I had meant that with regard to him and his situation. "No, no, the poor people who live there on the islands in the Pacific. That's bad," he replied. I asked him what he me ant by that. "Those are the victims of climate change. I heard the other day that the islands are going under, aren't they?"

I finished my conversation with Dirk, a farmer and central German victim of the climate crisis. Except he didn't make that connection.

If the geophysical reality pushes the boundaries of our imagination, a language is needed that does justice to this reality, present and future. We need climate translators who formulate the findings of science in such a way that people can connect them to their lives. It must become tangible what 1.5 or 2 or 3 degrees of average global warming mean for everyday life.

Instead, Hollywood producers, journalists, and ill-informed but vociferous media makers shape narratives that are reproduced again and again unquestioned. Nothing stands as exemplary for this as the narrative of the sinking island. It is true, of course: The flooded island is a thoroughly realistic scenario that should give us pause. The message it conveys is nevertheless grossly inadequate: The climate crisis appears here as the problem of a small number of defenseless people at the other end of the world. It moves away from us into the far distance. We are told to feel lucky that we are not a Pacific island nation. And yet we are supposed to get involved against the climate crisis? Well, fine, but that would be at best a charitable act.

A PROBLEM OF VIVIDNESS?

It has never been easier to get information than it is today. After all, we carry the accumulated knowledge of humanity around with us in our pockets. Sounds promising, but it doesn't work. Because this principally liberating possibility is accompanied by a feeling of being overwhelmed. More and more rapidly we get lost in the fog of information, and it becomes increasingly difficult to recognize false information and noninformation as such. Fake news has shaken the very sovereignty of knowledge. Who still dares to rely on a rock-solid truth when truths are suddenly being questioned everywhere and all the time?

The climate crisis is not only, but also, a crisis of communication. It starts with language. How can developments be put into words that have never happened before, and are partly beyond our imagination? We are dealing with a partially apocalyptic finding that requires a highly complex scientific analysis. What we are talking about is either microscopically small or macroscopically gigantic in magnitude, moreover mostly invisible and for that reason alone difficult to comprehend. The presentation of the climate crisis has a vividness problem. In addition, scientific literature is written in a language not designed for general comprehensibility. How do you resolve this and make the climate crisis understandable to a mass audience?

To this end, an entire branch of science has been developed over the past fifteen years: climate communication. This was urgently needed. For there is hardly any other current scientific field in which the overwhelming majority of scientists agree on the essential findings as much as in this one (climate change is real and human-made), which is nevertheless so aggressively doubted. The climate crisis is declared to be a question of faith—as if there were no objectively valid facts about it. It is absurd: We live in a time in which there is more reliable information than ever before, and yet a significant part of the public declares that the geophysical facts around us are a matter of opinion.

A psychological circumstance plays into the cards of those who promote such doubts: Scientific facts are not readily accepted by people as truth. They get filtered. If they correspond to one's understanding of values and worldview, they are accepted without further ado, and if not, they are not.

FRAMES INSTEAD OF FACTS

Facts are not always relevant for political decisions either. According to Elisabeth Wehling, a cognitive scientist who has been widely cited in recent years, it is frames, or frames of interpretation, that make the difference in political debates.[5] This approach looks at how consciously or subconsciously we experience things and how these impressions shape our actions. Frames, Wehling emphasizes, play a central role in this process. They are based on our physical, linguistic, and cultural experiences and "give meaning to individual words by placing them in a context with our knowledge of the world."[6]

Therefore, it has been rightly argued that it is incorrect to speak of "waves of refugees" in the migration movements of 2015. Whoever uses this term evokes images of people "flooding" the country in large numbers—just as waves do. In Germany, older people will also involuntarily think of the "flood of asylum seekers" when they hear "waves of refugees." Thirty years ago, this fighting term was used to stir up fears of masses of foreigners entering the country uncontrollably. Such charged terms strongly contribute to the way a debate develops, and to the interpretive frameworks employed to discuss migration movements.

Researcher George Lakoff, Wehling's doctoral thesis advisor at the University of California–Berkeley, points out another crucial aspect about frame research: Frames are linked to the emotional center of the brain. Each frame, once activated, triggers certain emotions in an unconscious way, according to Lakoff.[7] So when people talk about "climate change," they not only have images in their heads, they also feel something about these images. Many may be saddened or depressed, others may react unnerved or suspicious. The more often a frame is invoked, the more rigid it becomes in common usage and thinking.

If this theory is correct, it is of far-reaching significance for the analysis of political opinions and political action. In politics, says Wehling, frames usually function "ideologically selectively."[8] They focus attention on certain facts while disregarding others. Therefore, in politics, it is not only facts that guide action, but also the frames that create meaning.

If one follows Wehling's and Lakoff's argumentation, then political influence owes much to the ability of socially dominant frames to be able to influence the political process. With them, certain options for action move into the foreground, while others lose significance: This is a critical factor for the choice of words in successful climate communication.

We speak of a crisis because the terms "climate change" and "global warming" are unable to convey the danger of the current situation. And we are not alone in this choice of words. UN Secretary General António Guterres and the former director of the Potsdam Institute for Climate Impact Research (PIK), Hans Joachim Schellnhuber, also speak of the climate crisis. In May 2019, the English daily newspaper the *Guardian* declared that it was adjusting its editorial guidelines. Instead of "climate change," it would now use the terms "climate emergency," "climate crisis," or "climate breakdown." The term "global heating" should be preferred to "global warming." Emergency and crisis are terms appropriate in light of the facts; it also seems correct to speak of collapse in view of collapsing ecosystems.

Even in political arenas, we are slowly beginning to debate what language is appropriate to the current situation. What is at stake is nothing less than the interpretive sovereignty over the crisis by means of the terms that are used and the images and feelings that are activated. While some in German politics are still busy denying the existence of the crisis or downplaying it and others are bemoaning the "apocalypse whispering" (Harald Welzer) or the "total pessimism" (Bernhard Pörksen) of the debate, Elisabeth Wehling emphasizes (in the spirit of the *Guardian* and others) that the term "climate change" is inappropriate to capture the dramatic threat of global temperature rise. The abstract term "climate" puts the problem "far away from our everyday life," and the term "change" even allows a neutral reading—as if it were still unclear whether the situation would improve or worsen in the future.

According to Wehling, the consequences of the frame "global warming" are even more dramatic. Warmth has a positive connotation, and the everyday language term "warming" is linked on an emotional level with positive feelings ("warming up to something"). The term is therefore tantamount to a "cognitive happiness pill."[9]

That is why we speak of crisis. The ancient Greek word *krisis* means "decision," sometimes also "climax." We are in a crisis because the world in which we will live in the future will be decided in the next few years. If we have about seven years left to reach the 1.5-degree target and halve global emissions, policymakers are at a crossroads today. They can, indeed they must shape an era, as Abraham Lincoln, Franklin D. Roosevelt, or Willy Brandt did. A political era marked by a radical change of direction. Nothing less is needed for our future. And this can only work with concepts and images that convey this spirit.

CALCULATED UNCERTAINTY

We have talked about the fact that over the past fifty years, a debate has flared up about the climate crisis that is reminiscent of a religious war. That's not a coincidence, it's a calculation. The calculus of communication.

It is an open secret that energy giants like Exxon (now Exxon-Mobil, one of the world's largest oil companies) have known about the disastrous consequences of CO_2 emissions since the 1970s. Not only were they the first to know about it—Exxon even developed climate models themselves, the *Esso Atlantic* supertanker has been collecting extensive data on CO_2 concentrations in the oceans since 1979, and they employ some of the best experts in the field.

After the high-profile testimony of climate scientist James Hansen before Congress in June 1988, Exxon became concerned. That's why they joined with other oil and gas companies, power companies, and the coal and auto industries to deliberately spread doubt and misinformation about global warming for years through the so-called Global Climate Coalition. Even the US Chamber of Commerce was a member of this lobbying organization for a time. Using the same strategies that had successfully served the tobacco industry for decades to cover up the connection between nicotine and cancer has allowed talk of "uncertainty" to become entrenched in the political, media, and public mainstream. Emphasizing how unclear the situation is has been commonplace ever since.[10]

It is well researched how disastrous it was that the media in the United States were supposed to report on the climate crisis in a "balanced" way, which meant that dissenting opinions should be included in any reporting on the crisis.

Harvard historian of science Naomi Oreskes examined the scientific articles on climate change published between 1993 and 2003. She found not a single one among the 928 studies that questioned a human influence on global warming. Not a single one![11] This is why Oreskes also speaks of a "scientific consensus on climate change." A 2003 study of reporting by major US newspapers—the *New York Times, Wall Street Journal, Washington Post, Los Angeles Times*—showed that almost 53 percent of the articles quoted representatives of "both" sides. In another 35 percent of the texts, the human influence on the climate was emphasized, but the so-called skeptics also had their say.[12] That is how public perception of the state of scientific knowledge has been distorted since the 1990s.

Where did the voices come from that sowed doubts about the basic scientific consensus at that time? Among these experts was Frederick Seitz, who had been involved in the construction of the atomic bomb and was a past president of the US Academy of Sciences. William Nierenberg, on the other hand, had previously made a name for himself as a physicist through his participation in the construction of the atomic bomb, and as director of a renowned marine research institute. Finally, the astrophysicist Robert

Jastrow held a management position at NASA, where he was responsible for the US space program, among other things. All of them were highly respected physicists who had achieved some fame during the Cold War.

These three, Seitz, Nierenberg, and Jastrow, were also the ones who founded the George C. Marshall Institute in 1984 to support the Reagan administration's missile defense program. Even before the end of the Cold War, they began systematically attacking disagreeable research findings in concert with other think tanks, lobby groups, and political hardliners. They attacked studies on the cancer risk of smoking, the asbestos hazard, acid rain, the so-called ozone hole—and global warming. It worked brilliantly. The strategies and procedures of this "network of denial"[13] are well documented by numerous court cases, especially against the tobacco industry. This is also true for their influence on climate science, as Naomi Oreskes and her historian colleague Erik Conway describe. The professional deniers systematically cast doubt on the main findings of climate research, such as global warming: "At first, they claimed that it did not exist; later, they claimed that it was only a matter of natural fluctuations. Finally, they found that even if global warming existed, it was not that bad, and that we could simply adapt to it."[14]

The confusion created by this dangerous "balance" in reporting is still kept alive today. The #ExxonKnew campaign has long drawn attention to the extent of the influence these interest groups have on our perception and assessment of the climate crisis. To this day, institutions like the US-based Heartland Institute invest millions worldwide to downplay its explosive nature. This network is by no means an isolated case, but has been adopted over the years as a business model in all kinds of scientific subdomains.[15]

Germany, too, is familiar with institutions that deliberately interfere in the climate debate in order to push through their donors' agenda, for example the New Social Market Economy Initiative. This lobbying organization, financed by the employers' associations of the metal and electrical industries, tries to influence public perception through publications and advertising campaigns. The basic tenor of its current campaign is that yes, "climate change is currently the greatest challenge facing mankind," but efforts at the national level are not effective enough, are too expensive, damage Germany as a business location, and endanger jobs. Effective climate protection can only be achieved through innovations by industry, through the expansion of emissions trading, and through the decoupling of economic growth and resource consumption.[16]

In the interplay of right-wing populist and conservative politicians, influencers and inter-agency networks, political opponents and their environmental and climate agendas are systematically called into question. This has led to a situation that could not be more absurd: Of all people, those who even rudimentarily understand the drastic nature of the situation and infer the necessity of political measures from it, are labeled irrational ideologues, hysterics, or climate fanatics. That is why we speak of a crisis of communication when we speak of the climate crisis. The question of whether or not to face the reality of the climate crisis no longer depends on one's level of information, at least not solely. It has become a question of political framing.

BEYOND OUR IMAGINATION

Another communication problem has to do with the scale of the crisis: The climate crisis is too extreme. The truth about humanity's failure so far, about the extent of the destruction already being wrought, and the dangers yet to come—all this is difficult to grasp and hard to process. How does one react to such overwhelming messages? One fights against them or turns away. In psychology, this behavior in stressful situations is called the fight-or-flight response.

Those who fight it go on the offensive: They either question the message ("climate change is a lie") or at least the messenger of the message ("Greta has no idea," "Luisa has flown before"); they deny climate change or downplay how bad it is. Those who flee it turn away and leave the debate to others, feeling that they can't do anything about it anyway.

That is the problem: It is difficult to talk about this crisis. On the one hand, because of a decades-long, targeted disinformation campaign that has succeeded in making the crisis appear to be a scientifically disputed niche phenomenon. On the other hand, because the sheer intolerability of the truth often creates a feeling of powerlessness and resignation.

LUISA There are moments when everything feels pointless. When the battle seems to be lost before we have really begun. I guess that's called *weltschmerz*, world-pain in the truest sense of the word. June 18, 2019, was one of those days. I was making coffee while scrolling through the *Guardian* on my phone when a headline about tipping points stopped me in my tracks.

Almost exactly three years before, in the second semester of my geography studies, I had learned something about the so-called tipping points for the first time. Tipping points are the scientifically defined points at which unstoppable, irreversible chain reactions are triggered. In climate science, a tipping point is for example when the surface of the Greenland ice glacier melts or the Nordic permafrost thaws. From these moments on, we can no longer stop the warming. It takes on a life of its own, picks up speed, and transforms the earth into what scientists call a "hot house," a planetary sauna.

The message from the second semester was clear: The tipping points are the beginning of the end. They must be avoided at all costs. In almost every speech I've given since, I've talked about these disastrous thresholds.

So now I was standing in the kitchen in the morning reading the *Guardian*'s front-page story, "Scientists are shocked. The Arctic permafrost is thawing seventy years earlier than calculated."[17] A development that had been expected for the year 2090 at the earliest could already be proven now. Thus, a turning point had probably been reached. Another article on the thawing of the permafrost even spoke of a "catastrophic misjudgment"[18] on the part of science. It had long recognized the danger and foreseen this event, but was now itself surprised by the speed and severity of the crisis.

To many, the climate crisis seems a distant and abstract phenomenon. For me it is different. The climate crisis is part of my life; I see it in every aspect of everyday social life. And never before had it felt more threatening than that morning. For years I had been warned about this event. An event that was now suddenly here. It was as if the climate crisis had suddenly arrived in the middle of my kitchen. As if it had settled on my shoulders, never to be shaken off again. The news of the thawing permafrost left me stunned. And I was also stunned that no further headlines followed. There was no crisis meeting, no breaking news, no special broadcast. The day passed. As did the next. And the one after that. As if nothing special had happened.

It is easy to underestimate the relevance of tipping points. But at a time when the crisis has become the norm, the central challenge is to make the urgency clear without triggering resignation, which is why it is crucial to talk about tipping points. They must be understood. Because they tell of a climate system that is not gradually becoming more broken and dysfunctional. Not bit by bit, slowly and leisurely, until we manage to put a stop to it. The tipping points make clear that climate destruction can be irreversible in parts. The knowledge of this challenges us to put aside the human arrogance with which we imagine, in biblical fashion, that we can "rule" over creation. We must accept the realization that we cannot "repair" the climate unconditionally. Planetary rehabilitation potentials have their limits. It is like a kayak. Up to a certain tilt angle, a force acts to right it, but above a certain angle, the same force causes it to capsize. Tipping points are humankind's deadline, cynically not only in a metaphorical sense. It is possible to predict roughly when these tipping points will occur, and researchers such as Johan Rockström of the Potsdam Institute for Climate Impact Research have been able to discover a great deal about it. But the thawing permafrost has shown that everything can happen much faster. And that we must stop it under any circumstances.

For climate communication, the task is clear: start talking about timing and urgency. Because there is no such thing as slow climate protection—the kayak capsizes when the tipping points are reached. Consequently, communication must also address the timeline. It is often said that climate protection must be a combination of ecological with social and economic concerns. This triad needs a fourth dimension—time. Even the best climate protection projects will be in vain if they are not carried out within the time frame set by geophysics. This scientific truth is overwhelming and hard to bear, but it must be told objectively, accurately, and with appropriate amplification. Climate communication must be demanding and challenging. Many ask us: "Do we need fear? Do we need panic?" What questions! If you look at the data and take them seriously, of course you get scared. That is only human. The question is, what do we do about it? Communicating the climate crisis must create a constructive framework for the intolerable: for dealing with news that makes it difficult for anyone to see the future, especially those who still have so much of their lives ahead of them.

THE CLIMATE OF THE MEDIA

When we started working on this book, it seemed absurd to us how the debate was being conducted. In the media, the climate crisis was usually treated like a fertilizer regulation—it was a niche topic written about somewhere on the science pages. Sort of like, "It's worth talking about, but it also gets boring very quickly." Two years later, that has changed at least a little. Finally, the climate is making headlines.

The fact that the climate protests have become more prominent in the media has had an effect as well. More and more often, climate activists or scientists are invited to write for the politics or opinion pages or to speak at business conferences. But they still remain exotic guests, they are not presented as experts on a problem that people are concerned about. One cause of our current crisis of communication is that a large proportion of journalists have been out of the loop for too long. They report only when something spectacular happens; controversies that are as heated as possible get people excited and increase the number of clicks. Media companies above all should take responsibility for educating the disinformed public through good climate journalism.

There is still a lack of understanding that the climate is not a topic like any other. Media outlets that report more often and put the climate issue in the spotlight are quickly labeled as "activist." Apparently, it is a scandal to bring up an existentially important issue when there is currently no political row about the topic. The climate crisis demands other forms of media presentation. It is not enough to wait for a major disaster to happen in order to generate "news value." The climate crisis happens every day, all over the world. It is a permanent threat. If the media treated every melting glacier, every burning forest, every extinct species like an assassination attempt, the news would be full of them. After all, these are the many small disasters that, taken together, make up the climate crisis: They are assassination attempts on humanity.

But when there are headlines, they are too often produced by political journalists without scientific expertise or even basic knowledge. Thus, the climate ignorance of political decision makers is at least indirectly supported by the fact that there is no critical questioning of current climate plans and no exposure of broken climate policy promises.

The (conservative) political establishment benefits enormously from the fact that the coverage of the climate crisis is often so terse. There is a lack of voices challenging politicians, denouncing their inaction, exposing their inexcusable lack of a plan. Until Fridays For Future took to the streets, there was almost no confrontation. This movement collided full force with policy-makers who were shockingly clueless. Where do you start when members of parliament, ministers, mayors—predominantly men, by the way—react to the central question of the century with a shrug of the shoulders? When they really hardly know how to implement the Paris Agreement politically? When they can't explain the concrete consequences of exceeding the 1.5-degree target, and what would have to be done to meet this target?

A new study published in August 2019 highlights another disturbing trend: A comparison of the media presence of climate deniers and scientists found that deniers were mentioned 40 percent more often in the media than experts.[19] It is important to note the formats that were examined, because in traditional media, the difference was only 1 percent. This is because the overwhelming majority of baseless statements can be found in new media formats, whose large presence carries the risk of displacing serious journalism—and along with it any factual reporting on the climate crisis.

It is also up to the scientific community to educate decision makers. Scientists, however, are too seldom consulted on the topic; for too long, there has been a lack of societal pressure. Moreover, scientists are not experts in communication and hardly have the budget for effective public relations. Exceptions confirm the rule: In Germany, the German Advisory Council on Global Change (WGBU) fills the gap, at least in part. But who ensures that the proposals of science are actually followed? This, too, is a question of public and economic pressure. But where this pressure is to come from remains unknown as long as politicians and the media do not convey the seriousness of the situation.

How do we put something on the media's agenda that concerns us because it threatens our civilization in an unprecedented way? Even while the ugliest images produced by the climate crisis are still coming from faraway places? Our journalists are far from doing their homework.

Who acts when no one speaks? Who acts when people speak but no one listens? This is where climate communication goes around in circles. What remains is an uninformed or even misinformed and sometimes resigned public, distracted politicians, and a media world that pays more attention to every soccer match than to our ecosystems, which are collapsing by the dozens.

HOW DO WE GET OUT OF IT?

By telling the truth. In the appropriate language. By telling stories that reach people. We need a language that makes the crisis tangible in everyday life, not in a paralyzing way, but in an activating way. Are we trying to scare people with this? We can't change it: The reality of the crisis is, well, scary. But we want fear to be constructive and awaken each individual's sense of responsibility and willingness to act. Quite in the sense of Günther Anders, who, under the impression of a possible nuclear war, called for the "courage to fear." We need, as the philosopher put it: "1. a fearless fear, since it excludes any fear of those who might mock us as fear mongers. 2. an invigorating fear, since it should drive us out into the streets instead of into the corners of our rooms. 3. a loving fear, which should fear for the world, not only for what might happen to us."[20]

Fearless, invigorating, loving. Perhaps it is these qualities that the climate crisis and its attendant communication crisis require of us today.

7 THE CLIMATE CRISIS IS A CRISIS OF FOSSIL CAPITALISM

Our parents were told that sustainability had to be compatible with economic growth. The credo was that market forces were the guarantor of a green and prosperous future. Decades later, this has hardly changed. Like our parents, we are told that ecology, economy, and social welfare are the three pillars of a forward-looking policy. That sounds coherent, a beautiful narrative. But it doesn't add up in reality. After decades of growth, the introduction of market-based instruments, voluntary commitments by industry, and publicly financed incentives for companies, "market-based democracy," as Angela Merkel called it in a speech in 2011, is heading straight for climate collapse.

More and more urgently, the question arises as to how long we should continue to hope that the invisible hand of the market will get a grip on the climate crisis and conjure up a CO_2 price out of the hat that is high enough so that the actual costs of each ton of CO_2 are covered. That the market will ensure that the Paris climate targets are met. That the market will spit out the one ingenious invention that will save us from climate chaos.

Chances are high that none of this will occur in time, as long as we rely on "the market" to sort it out for us. The past decades have shown that voluntary commitments are not enough if the basis of life is to be preserved for future generations. So why wait for that? We have known for a long time that the way we do business, produce, and consume pushes the planetary boundaries.

Strictly speaking, these are new limits, which Johan Rockström, director of the Potsdam Institute for Climate Impact Research, defines as follows: Climate change, ocean acidification, ozone depletion, phosphorus and nitrogen cycling, global freshwater consumption, land-use change, biodiversity loss, atmospheric aerosol density, and chemical pollution.[1] For the past 10,000 years, these systems have been largely stable, but now they are out of balance. Species extinction most dramatically so, according to

Rockström. According to the United Nations World Biodiversity Council, nearly a quarter of land areas are now ecologically degraded and no longer usable. About 85 percent of wetlands have already been destroyed, and 9 percent of all livestock breeds are extinct. One million species are threatened with extinction in the coming years and decades. The rate of species extinction is thus at least ten times higher than the average of the last ten million years. The trend is upward.[2] According to Rockström, the nitrogen cycle and climate change are also already damaging species to an extent that they threaten the livelihoods of wildlife and humanity.

It is actually quite obvious. If an understanding of the existence of planetary ecological boundaries is not enough, their transgression must be legally halted; if the massive use of nitrogen and phosphorus in industrial agriculture destroys the fertility of soils, we must limit their use; if carbon dioxide heats the earth, we must take the biggest CO_2 emitters off the grid and ban them from the roads and the air; if methane has a similar effect, we must limit meat production; if emissions in inner cities make children and the elderly sick, they must be reduced; if exorbitant use of nitrates contaminates our drinking water, this practice must be regulated.

None of this is an end unto itself or even an expression of "hypermorality," as some claim. It is simply a matter of drawing consequences from the realization that we have reached a historical caesura: The narrative of win-win situations has turned out to be a myth; it has led to a dead end. In order to stay within the planetary boundaries and within the framework of the Paris Agreement, an energetic restructuring of infrastructures is needed, and this will not be possible without cuts in our climate-damaging habits: be it for transport, energy production, industry, or agriculture. This will not happen "voluntarily" or even be brought about by the "market" but must be enforced with regulatory measures, bans, and incentives.

It is a widespread misunderstanding that bans are always radical, that they always stem from left-wing politics, and that they inevitably take something away from people. Conservatives and liberals therefore hurry to distance themselves from them. Angela Merkel regularly emphasized that she did not want to govern "by bans and regulations," but rather by "innovation."[3]

The at times hysterical reactions to proposals for state regulations and bans clearly show how dominant the belief in the salutary powers of the market is in Germany. The categorical rejection of binding sectoral targets for emissions, for speed limits on highways or driving bans in inner cities, is a symptom of the deep roots of neoliberal thinking in our political system. What is radical about banning business practices that are harmful to the environment, climate, and health? It is radical *not* to regulate them. What is radical is to instead create incentives to provide tax relief to companies that harm the environment, climate, and health. It is radical to defend a continuing ecological madness in the name of freedom. Radical, in the most destructive way imaginable.

THE FATEFUL BELIEF IN THE MARKET

Friedrich August von Hayek, a central founding figure of neoliberalism, once described it as a "fatal conceit" of humanity to want to control economic processes better than the principle of supply and demand can.[4] Hayek considered the market economy to be the highest stage of a cultural development that resulted from the transmission of cultural and genetic behavior patterns. As a result of this quasi-evolution, the market was an order of magnitude superior to other reason-based forms of social order, in his view. Therefore, Hayek urged, we should let ourselves be guided by the invisible hand of the market and firmly trust that there is no more rational principle to organize the national economy.

We live in a society today that makes many political decisions based on this belief. Too often, "the market" is blindly defended, and too seldom is the question raised as to what this "market" actually is.

Whoever hears "market" may think of the weekly market around the corner; a place where people offer fruit, vegetables, cheese, etc., for sale. Depending on how many people want to buy their apples or avocados here, and also depending on what the prices are at the neighboring stand, the sellers shape the price of their offer, so the lesson of the market goes. The lower the price or the scarcer a sought-after good, the greater the demand. And vice versa. In this way, prices fluctuate continuously, so that a natural balance of supply and demand is eventually achieved. But this ideal picture is false even for the simple example of the weekly market. In Europe every kilo of potatoes, every bunch of radishes, is subsidized by the government. And could it be that not only the invisible hand of the market, but also the silent surplus income of the traders warrants that the price for a head of lettuce does not fall below a certain amount?

The ideal image of the market makes it easy for the representatives of so-called market fundamentalism to anchor the theory of a self-regulating system of supply and demand in political discourse. We speak of market fundamentalism here because that is the vanishing point of any proposal from this direction: the more market, the better. Especially in economic and social policy, but also in climate policy, "the market" enjoys a mostly unquestioned reputation as a bubbling source of seemingly efficient and economically compatible policy approaches: fundamentalism at its best.[5]

The belief in the market as the highest but undefined good takes on forms that could not be more absurd. Economist Stephan Pühringer has analyzed Angela Merkel's speeches from 2008 and 2014. He shows that she repeatedly speaks of "the market" as if it were a person who must not be "alarmed" or must be "persuaded" by political ideas. A person with vital "powers" that must be protected by politics, since under no circumstances should politics interfere with the self-regulating mechanisms of the market.

In this model the market behaves like a bar of soap: As soon as you try to grab it, it slips out of your hand. As the economist and cultural historian Walter Ötsch explains, the term is loaded with different, sometimes contradictory meanings and can be used flexibly, depending on political concerns. Sometimes it is understood as a real phenomenon expressed in institutions or laws, sometimes as a norm to be followed, sometimes as fiction, potentiality, or utopia, i.e., as something worth striving for but still awaiting its realization.[6] Practically, this manifold way of usage immunizes against criticism: Depending on the need, it assumes a different meaning.

What is usually withheld in these debates: The forefather of economics, Adam Smith, never spoke of "the market" as a singular, abstract principle; when he spoke of markets, he always meant concrete places of trade and exchange. The concept of the market as we use it today was developed in the 1920s by the liberal economist Ludwig von Mises. His student von Hayek developed it further. Throughout his life, von Hayek tried to give the market principle a philosophical justification.

While Adam Smith always understood the market as something concrete, von Mises, von Hayek, and their successors have elevated it to a general principle of order, which is used as an argument against almost every form of government regulation. In this context, the market is constantly connoted with positive terms such as freedom, reason, and progress, while the state is associated with negative aspects such as coercion, burdens, and distortion of competition.

In part, this dichotomous thinking can be explained by the historical context of the 1920s and 1930s, which was strongly influenced by the competition of systems. Under the impact of Lenin and the socialist-influenced "red Vienna" in which von Mises lived, he wrote in 1929: "There is no other choice than this: either to refrain from isolated interventions in the play of the market or to transfer the entire management of production and distribution to the authorities. Either capitalism or socialism; there is no middle ground."[7]

Today, ninety years later, the world is a different place, and perhaps von Mises would judge it differently as well. But the ideas about the role of "the market" for economic development, which he and von Hayek shaped in the most profound ways, continue to persist.

LUISA There are conversations that are so similar in structure and progress that one could write essays about it. I have been part of many such conversations. No matter where they take place—whether on small, leather-covered swivel chairs in the studio of the talk show *Hart aber Fair*,[8] in bustling newspaper editorial offices in Berlin-Kreuzberg, on podiums under dim spotlights, in cafés during "backroom talks," or between doorsteps at a reception—our conversations—by us we mean those who want to put the climate on the agenda—with those who have set this agenda in recent decades, i.e., news editors, politicians, and entrepreneurs, usually men in suits, follow the same pattern.

As a rule, the climate crisis was typically not on these people's agenda in the past. At best, they now half-heartedly admit that they "have to do something in view of the climate." (But there is no hurry.) Now that they have suddenly been challenged by Fridays For Future, however, they are suddenly asked to have an opinion on the subject. These conversations can be divided into three phases. They often begin with encouragement or recognition. Very polite and respectful, of course. That's followed by a few minutes of "you should": You should make more of an effort, you should have a serious word with China, you should also look much more at Africa. And then, after it has been stated that climate protection is very important, the big But follows. A "but" that cannot be avoided, a "kill-but": But the market will take care of it, because the market can do it. The invisible hand of the market will tackle the problems for us, it will look for and find the best means. The market will show us the way to achieve the climate targets, leaving us our "prosperity," our "growth," our industry, and our jobs.

That's what Herbert Diess, the CEO of Volkswagen, means when he talks enthusiastically about the market launch of electric cars on talk shows; that's what Finance Minister Christian Lindner means when he talks about innovation and market power wherever he goes; that's what the editorials in *Die Welt* and the *Frankfurter Allgemeine Zeitung* mean when they demonize "bans" without differentiation and warn of paternalism and recession. They are all agenda setters. They paint the paradoxical picture of a market that, on the one hand, is portrayed as a superhero who can save our climate, but, on the other hand, seems to be a fragile and irritable little man that must not be angered or worse, shaken.

So, before we dare change the human-made system of "the market," we opt for changing the natural system of the world's climate. As if we were more concerned about the well-being of the market than the future of our civilization. These conversations always leave me convinced that the end of the climate crisis will remain unattainable if we do not study this fear.

In much of the world, neoliberal-style fossil capitalism has become the dominant principle. In many places, it threatens the livelihoods of human life through excessive resource consumption. Ironically, "market-based instruments" at the national and international level are still considered the only promising answer to this threat.

Since the 1980s, parallel to the rise of neoliberalism, which became a systemic shooting star in the United States under Ronald Reagan and in Great Britain under Margaret Thatcher, so-called market-based environmental protection has become increasingly important. In the face of increasing deregulation, weakened governments, privatization, and government budget cuts, it was left to markets to protect and manage nature. At the heart of this approach was the allocation of title deeds to land, water, forests, and fisheries, and private sector trade in these resources. The assumption was that through high prices the "free market" would create incentives for their sustainable management and use and thus contribute more efficiently to environmental protection than government ownership of natural resources could.

At least by the 1990s, this approach also became relevant for dealing with the climate crisis. Since the Kyoto Protocol of 1997, trading in CO_2 certificates and other "pollution rights" has been the central approach to reducing greenhouse gases worldwide. The European Union's Emissions Trading Scheme, modeled on the Kyoto Protocol, has created the world's largest emissions market. China launched the world's largest national emissions trading system in 2021, while Germany introduced a national fuel-trading scheme the same year. More than twenty such trading systems have already been established worldwide, and countries such as Indonesia, Colombia, and Chile are currently discussing or testing their introduction.[9] In 2021, however, only around 21 percent of global greenhouse gases were traded on such exchanges.[10]

A PRICE TAG ON NATURE IS SUPPOSED TO SAVE US. SERIOUSLY?

The steady expansion of emissions trading reveals how paradoxical the situation is: On the one hand, this instrument is still considered the best approach to climate policy by many of those in charge in politics and business. After all, in theory this approach is supposed to be extremely efficient. On the other hand, its effectiveness is still surprisingly low. Global emissions have also risen steadily over the last two decades, and this is not only due to the Chinese coal boom. And the fact that there has been a slight decrease in emissions at the European Union level in recent years is mainly due to the expansion of renewable energies[11] and only in part to pollution rights trading.

The reason for the moderate results of emissions trading is not the ineffectiveness of the principle itself, but the price of the certificates, which is much too low. That is why more and more regulations and increased interventions in the markets have been observed recently. The expansion of emissions trading on the one hand and its stronger control on the other are intended to improve its effectiveness.

What is clear, however, is this: There is little reason to believe that our future is in good hands with the market and its emissions-trading instrument; on the contrary, it would be foolish to entrust it with our fate. We must recognize that after decades of focusing on market-based climate protection measures, we are moving with great strides toward irreversible tipping points in our global ecosystem, and we are running out of time.

That is not all. For it is doubtful whether the market principle as such is suitable for solving the climate crisis. For the economic historian Karl Polanyi, who was a kind of counterpart to von Mises and von Hayek in Vienna in the 1920s, the cause of the crisis lay in the market economy itself. From the beginning of the industrialization that began in the nineteenth century, Polanyi observed how ever-larger fields of society were subjected to the logic of the market principle until this logic fully became the "matrix" of the social system. Society in its entirety became nothing but an appendage of the market.

"Such an institution could not exist for any length of time without annihilating the human and natural substance of society; it would have physically destroyed man and transformed his surroundings into a wilderness,"[12] Polanyi warned in his 1944 book *The Great Transformation*. In it, he outlined the genesis of the market economy and warned of its destructive consequences.

Polanyi grasped early on what is still true today: Since markets do not function without commodities and prices, ever-greater parts of our social reality are gradually becoming tradable commodities. An absurdity that we perhaps hardly notice anymore only because it has become all too normal. Our descendants may well look back on our actions one day with as much incomprehension as we look back today on the slave markets of the eighteenth century: We stick a price tag on everything that surrounds us. Not only do we trade in raw materials and the goods produced from them, in animals, real estate, financial products, bets on fluctuations in the stock markets, and insurance for and against everything and anything—we also trade in a right that we ourselves invented in the first place: the right to pollute our planet, the atmosphere, and our fellow human beings.

The transformation of nature into a "fictitious commodity," as Polanyi called it, was for him "perhaps the most absurd undertaking of our ancestors."[13] For him, land was and always remained nature, inextricably interwoven with the circumstances of human life. The soil "gives steadiness to man's life, it is the place of his dwelling, it is a condition of his physical security, it signifies landscape and seasons."[14]

What Polanyi reminds us of is the simple fact that we are only guests on this planet. And the way we are currently treating our host, it is hardly surprising that it is making life difficult for us. The road to doom, it seems, is paved with well-meaning, market-based instruments.

ALEX When I was fourteen, a friend of my father's gave me a book by the architect and futurist Buckminster Fuller. The book was already falling apart, but with its yellowed pages it held a strange fascination for me. It was a collection of lectures from the 1960s in which Fuller spoke about the dangers and opportunities of technologies of the time for future life on the planet. He warned strongly against relying on fossil fuels as an energy source. What, he worried, would his grandchildren's generation use to power the machines when the supplies of coal and oil eventually ran out?

I understood immediately that the "generation of his grandchildren" meant me. But today, more than fifty years after Fuller's reflections, we have other concerns. When we have exhausted the reserves of fossil fuels, the planet will be uninhabitable for most people. Fuller's foresight is astounding to me from today's perspective, even if preventing climate collapse is now the motivation for a transition to the post-fossil age, not the fear of running out of fossil fuels.

I envy Fuller for the confidence with which he understood technological development as an engine for prosperity and progress. Fuller was convinced that technological development was the key to averting the looming dangers of war and resource scarcity. Neither politics nor any ideology could save us, but only what he called a scientific design revolution. Technology, Fuller hoped, could lead to an increase in the quality of life worldwide, making the utopia of a life of prosperity and peace for all a reality.

I am struck by the power with which Fuller spoke about the future. It manifested itself in his work: His futuristic city designs are legendary. A huge dome over Manhattan, providing clean air and a pleasant climate. Floating cities in the bays off San Francisco or Tokyo, seaworthy and shaped like tetrahedrons. In all this there is a visionary view of the world of tomorrow, which confronts the dangers ahead with technological sophistication, always confident that it can stay one step ahead of nature. Today, more than ever, this confidence has been shaken. The power craze of people like Buckminster Fuller, as can still be found today in Silicon Valley or in many approaches to geoengineering, helped to bring us into the crisis whose beginnings we are currently experiencing. As much as Fuller's farsightedness fascinated me at the time, today I am frightened by the arrogant belief that we can cheat nature through technological development instead of paying attention to the preservation of natural balances and cycles.

THE FIRST TIME AS TRAGEDY, THE SECOND TIME AS FARCE

In the early 1980s, the climate crisis was a hotly debated topic in the United States. There was widespread agreement that the greenhouse effect posed a threat to future generations. People envisioned scenarios of an ice-free Arctic and sunken port cities, and there was speculation about trade wars and political unrest triggered by climate change.

It was ultimately the *Changing Climate* report of 1983, a report involving leading scientists such as climatologist Roger Revelle and economist William Nordhaus, that proved to be an outlet for the ever-increasing political pressure to act. Revelle was one of the first to describe the greenhouse effect. As early as 1957, together with the chemist Hans Suess, he had drawn the conclusion that is today one of the most quoted statements in the history of climate research: "Human beings are now carrying out a large-scale geophysical experiment of a kind that could not have happened in the past nor be reproduced in the future."[15]

But twenty-six years later, the same scientists who had warned about global warming until then were all of a sudden telling everyone not to panic. In their 1983 report, they reiterated their deep concern about climate change. They also argued for an accelerated transition to renewable energy. But their conclusion was one of reassurance: While caution was warranted, further research was needed, but nothing more. Instead of making clear to politicians the urgency of preventive measures, the scientists advised them to hope for the ingenuity of the next generation, which would save itself from the impending disaster.

Today's technological solutions, however, are mostly nothing more than megalomaniac fantasies of researchers who want to blow aerosols into the stratosphere, accelerate global ocean circulation, or fertilize the oceans.

Research on "geoengineering," as these approaches are collectively called, is important for exploring ways to mitigate the greenhouse effect and its consequences. But there is no prospect of technologies that can even begin to contribute to a noticeable and controlled mitigation of global warming on the scale that would be necessary while being financially feasible.

What about nuclear power? Wouldn't that be a simple and proven alternative for energy production? The short answer is: no. Opinions are divided on this, especially in Germany where we have had a particularly energetic antinuclear movement, but also a long struggle to phase out nuclear energy. The extreme dangers to people and the environment, which technological developments have never been able to satisfactorily mitigate, are set against the advantages of an energy source that is costly but relatively low in CO_2.

In the second chapter of the Intergovernmental Panel on Climate Change's 1.5-degree report, published in 2018,[16] the situation is assessed as follows: There are scenarios for achieving the 1.5-degree target in which nuclear power is expanded against the current trend. However, it is also possible to follow the 1.5-degree path without this expansion. In this context, the IPCC emphasizes the major risks of the technology and the fact that investments in nuclear power are currently not profitable.[17] There are two factors that will determine whether the share of nuclear energy in global energy generation will continue to decline or whether the trend will be reversed: first, the way in which this energy transition is discussed in public, and secondly, the willingness to invest in the expansion and development of renewable energies. Furthermore, it is clear that Germany can achieve its own climate targets and also the targets agreed in the Paris Agreement without nuclear power.

Individual aspects of market-based approaches can certainly be part of the solution, but the central question is: Who will initiate the key changes needed within a few years to prevent the worst from happening? Every IPCC report in recent years has pointed to the need for enormous structural changes in our energy systems, infrastructures, land use, food production, and the way people and goods move from place to place. Such a transformation of society must be planned, coordinated, led, and vigorously pursued. No invisible hand can take this away from us.

It is a question of political will. Just as it is a decision to place economic growth above the rights of future generations, it is a decision to move in the direction of implementing consistent climate protection in the sense of intergenerational justice.

Hoping that the market or technology will deliver us from this existential crisis is reminiscent of what Karl Marx said about the course of history: "Hegel remarked somewhere that all great world-historical facts and persons occur, as it were, twice. He forgot to add: the first time as tragedy, the second time as farce."[18] While the "caution, but no panic" attitude of the 1983 *Changing Climate* report may, in retrospect, pass for a tragic aberration, the timid climate policy at the national and international level is in many ways a farce today, in light of the overwhelming scientific evidence.

There is another way. For the market does not have to be regarded as the authority that literally "holds our future in its hands" for all eternity. In May 2019, New Zealand's Prime Minister Jacinda Ardern said goodbye to growth and productivity as measures of government success. She announced that all new government spending must contribute to the following five priorities: improving mental health, reducing child poverty, addressing Maori and Pacific Island Indigenous inequality, developing the digital age, and transitioning to a sustainable, low-emission economy. Two weeks earlier, the government also introduced draft legislation with the goal of reducing New Zealand's carbon emissions to net zero by 2050.

There won't be the one principle that gets us out of this misery. Nor is it our intention to demonize across-the-board approaches that create incentives for companies, organizations, authorities, and individuals to act sustainably. We in Germany in particular have seen how the introduction of the EEG levy in 2000[19] temporarily revolutionized the renewable energy electricity market. We also know that a radical rebuilding of social structures must always take place step by step. And that it must begin where we are. But we should not be blinded by ostensible constraints and the attempts of political parties to sell us their unchanging agendas, decorated here and there by the term "climate," as new concepts for the future.

The market and technology will not save us, we have to do that ourselves. The overcoming of market dependence would not be the overcoming of democratic decision-making mechanisms, as so often feared. Rather, it would be a democratic setting of priorities oriented toward the well-being of society and the environment that redesigns human-made systems such as "the market" in people-friendly ways.

8 THE CLIMATE CRISIS IS A CRISIS OF PROSPERITY

LUISA The moment a complete stranger called and told me to immediately turn off my cell phone and buy a new SIM card after our conversation, I accidentally ran in front of a bus. Since the bus driver was more attentive than I was and stepped on the brakes fast enough, I was able to experience what it feels like to be in the center of a so-called "shitstorm."[1]

I learned that the caller worked for a political foundation and followed social media campaigns from the right-wing populist spectrum. A few hours earlier, a columnist who wrote for one of the country's largest media outlets, among others, had uploaded links to twenty photos from my Instagram account to Twitter and presented them, along with short texts, to his more than twenty thousand followers. The caller warned me that the next few hours would probably be a bit unpleasant. The photos showed me abroad, and I had flown to some of the places. The idea of using these photos to thematize my flight behavior worked brilliantly. Only a few hours later, my feed was flooded with criticism and hate messages: double standards, practice what you preach, hypocrite. Two days later, someone uploaded a video clip, and it was even accompanied by music. That, too, caught on.

A climate activist who had opted to fly more than once, that wasn't okay. The video was clicked tens of thousands of times, the media picked up on the topic, and Wikipedia added a "controversy" section to the article about me.

Finally, a document with my phone number was linked on a Twitter account.

The first time I was called by a man who obviously had my number from Twitter, I was almost disgusted by my cell phone afterward. Half an hour later, I was on live TV, trying to make myself clear.

It felt hard to motivate people to go to a strike at that point. I felt like I was deliberately exposing them to hatred against climate activists. I myself felt exposed, because the hateful voices on my cell phone display didn't let up for days, and I could never completely ignore them. It was strange to see dozens, sometimes hundreds of people talking about how little they thought of me under each post about me. And it was almost unbearable that people who stood by me, who spoke out for me, also became a target of hate and agitation on the web. But I kept going. Maybe even a little louder.

Much about this debate felt absurd. No matter what people wanted to criticize about me and my lifestyle, to what extent did this affect the plausibility of my demands? I can understand that people want to know how climate-friendly my lifestyle is. But that was never the point. It was an attack on me as a person in the hope of weakening a movement. But what concerned me most in those days was something else that was also resonating in the accusations against me: the question of whether a life in well-off circumstances is compatible with claiming to take responsibility and to look for ways out of the crisis.

Up until the point when this shitstorm started, I had never spoken out aggressively about the flight issue. Nevertheless, voices were now being raised from all sides accusing me of living a life that I forbid others to live. In the eyes of many, it seemed to be a logical conclusion: Climate protection means taking something away from people. Exactly what seemed almost secondary. What mattered was that "prosperity" was supposedly being threatened all at once—by people like me.

In the midst of this experience, which I wouldn't wish on anyone, I was most fascinated by this question: What does it mean for an affluent society when its prosperity becomes increasingly incompatible with people's responsible treatment of the planet and future generations? What happens when society becomes aware of this?

Yes, the climate crisis is a prosperity crisis. A prosperity crisis and a justice crisis that affects everyone. We have to talk about that.

BUT WE ARE DOING SO WELL, AREN'T WE?

At first glance, it seems that things have never been better: Extreme poverty has been halved worldwide in the last two decades, global life expectancy is at an all-time high in over seventy-two years, 80 percent of one-year-old children are now vaccinated against dangerous diseases, and 80 percent of people have access to electricity.[2] These are great achievements. The relative prosperity of the world can also be measured—traditionally in terms of global gross domestic product (GDP). It has never been higher than it is today; according to the International Monetary Fund, it was more than $84 trillion in 2018.[3] For economic policymakers around the world, investors, and businesses, global GDP is the pulse of the global economy. When GDP grows, economic strength is high and people are doing well.

According to Hans Rosling, former professor of global health in Stockholm, the world is getting a little better every year, bit by bit.[4] Yet most people have the impression that things are changing for the worse. Rosling, a kind of missionary of a fact-based worldview, called this the "negativity instinct."[5] He warned us against comparing extremes or averages, mentally extending current trends into the future, using large numbers without comparative values, or generalizing things unnecessarily. Only then would we be able to better identify developments that are actually worrying and take countermeasures.

Global greenhouse gas emissions, melting ice, and rising sea levels are among such worrisome developments, according to Rosling. The drama of the current situation becomes especially clear when one looks a little closer at what is causing global GDP to grow: 87 percent of the world's energy was derived from oil, coal, and gas in 2016.[6] And although the dangers of CO_2 in the atmosphere have been known for decades, six of the world's top ten companies by revenue in 2015 made their money from the extraction or sale of oil, coal, and gas.[7] The other four—Walmart, Toyota, Volkswagen, and Apple—are symbols for the unleashed individual consumption of our day and the status gain that comes with it. Never before has so much been bought so cheaply as today, never before have so many cars been on the road, and never before have so many smartphones been in circulation.

Those who speak of material prosperity are not only talking about an unprecedented availability of consumer goods, a unique variety of products, and a Western world that hardly knows scarcity anymore. This prosperity is based on fossil fuels and mass consumption. It promises a world of eternal availability and endless possibilities. Its motto: You only live once.

According to the economist Martin Kolmar, this type of prosperity is characterized by the fact that it no longer serves the satisfaction of basic needs, but primarily the satisfaction of needs for self-realization and status. These, however, are not only never to be satisfied, but they also constantly create new needs. Thus, luxury becomes a mass commodity. We are trapped in a "resource-burning machine." Nevertheless, this kind of prosperity is considered the epitome of the good life. But not all prosperity is the same. For instead of talking about prosperity measured in terms of satisfaction, happiness, health, or freedom, economic policy defines prosperity primarily in terms of GDP. Even to speak of "prosperity" is confusing—the term suggests a general feeling of well-being—as if a certain form of economic activity were identical with a general good mood.

This used to be different. When GDP was introduced as a category of measurement after World War II, it was based on the following theory: If GDP, i.e., the total production of goods and services, increases, the overall well-being of society also increases. The greater the purchasing power, the higher the tax revenues—and the greater the investment in education, social services, and jobs. With a growing middle class, most people would be better off. That was the promise.

There are only two catches.

One: The hope of growing prosperity across the board has not been fulfilled, at least not permanently. After the period of the so-called economic miracle after World War II, which was accompanied by an improvement in the standard of living for large sections of society, growing inequality has again been observed in many countries and regions in recent decades. While GDP is rising, wealth is being distributed to fewer and fewer people: In 2015, the richest 1 percent of the world's population owned more than the remaining 99 percent combined.[8] If this development continues, Kolmar warns, states will increasingly turn into de facto oligarchies whose basic democratic orders will be worn down by financial concentrations.[9] At the same time, the discrepancy between GDP growth and welfare growth is increasing not only in Germany. The latter is still at the level of the 1990s, while GDP has grown by 40 percent since then, according to a 2018 study by the Hans Böckler Foundation.[10] All this makes people unhappy.

In parallel, multinational megacorporations gained increasing influence in recent decades. Of the hundred largest economic players in the world (measured by their income in 2017), sixty-nine are now multinational corporations.[11] States are quickly becoming dependent on these players, and the consequences are competitions for locational advantages, which translate into low taxes for companies, for example. The first to be burdened are the citizens. To summarize once again: Promises of prosperity have not been kept in the past decades.[12]

WE ARE LIVING AT THE EXPENSE OF OTHERS

The second catch: This prosperity, which practically defines the Global North, is not self-sustaining. It is based on exploitation: the exploitation of social and ecological resources. Political scientists Ulrich Brand and Markus Wissen therefore refer to this prosperity as an "imperial way of living."[13] By this they mean the habits and structures of global capitalism through which Western industrialized regions appropriate the resources of the Global South and future generations. The term also refers to the fact that our everyday actions are closely interwoven with the social structures in which we live.

This raises uncomfortable questions about our way of life: Who sews our clothes? How much do those people earn? Who takes care of the ecological costs of the pesticide-ridden cotton monocultures or the chemical pollution of waters around the factories where jeans and T-shirts are dyed? Who picks our mangoes, and what impact does their cultivation have on local ecosystems? What happens to our old cell phone after we get a new one? Who are the people who sort our electronic waste in Ghana? How do they live? And of course: Under what conditions are the pigs kept whose ribs we like to eat so much? Where does their feed grow, and how is it traded? How do such production chains work? Who dominates them? In the end, all this ecological and social damage inevitably leads us back to the question: Who will repair the climate, which we are so recklessly destroying—and with it the prospect that genuine global well-being can develop sustainably?

The balance today is sobering: The higher a country's GDP, the more greenhouse gases it emits, statistically speaking. For a while, people hoped that this correlation could be changed, that global emissions could be reduced despite rising GDP. Those were the years after the financial crisis. In 2017, however, global emissions shot up again and increased once more in 2018.[14] Even if more and more countries manage to decouple the growth of their GDP from the growth of emissions, the global trend is clear: The higher a country's material wealth and thus its GDP, the larger its carbon footprint.

If we engage with the idea that the material prosperity that has unfolded to date in Western nations is burdening many people around the globe, making relatively few people truly happy, and imposing ecological costs that are so high that they have accumulated over decades into an ecological crisis, we find that the climate crisis is, at its core, a prosperity crisis.

At the end of the day, the population of the Global North will be asked to drastically reduce their resource consumption and radically change their way of life. Instead of claiming resources for themselves that would require several planets, as is the case today, they must live in a way that is compatible with the finite nature of the available raw materials and with human rights, including in other regions of the world. That is the demand. And the clock is ticking. While the consumption of the Global North alone is already using up disproportionate amounts of resources, the Global South is now following in our footsteps.

Unfortunately, a large part of the deprivileged population around the globe is following the wastefulness of the Global North. The American way of life is an export hit. What was long a way of living exclusive to the Global North is gaining a foothold in large parts of China, India, and South Africa. Just at the moment when its devastating effects have become impossible to ignore, our ideal of prosperity is spreading all over the world. Right now, the "generalization of the non-generalizable," as Brand and Wissen put it, is being promoted with great enthusiasm. And the multiple crisis we are experiencing today is an expression of this contradiction.

To break away from a GDP-obsessed ideal of prosperity without a future, we need new parameters with which to measure actual well-being. GDP can no longer be the measure of all things. And there are already a variety of such alternative indices. For example, the Global Happiness Index, which assesses life fulfillment in terms of mental health, happiness, well-being, and actualized values. Between 2016 and 2018, Finland, Denmark, and Norway were in the top positions, while Germany was only in seventeenth place, after countries such as Costa Rica and Israel.

The comparatively progressive Human Development Index also includes factors such as education and life expectancy. Here, Ireland and Australia, among others, rank ahead of Germany (in fifth place). Finally, the Kingdom of Bhutan is famous for being the only country in the world that no longer uses gross domestic product as its main development indicator, but rather the so-called "gross national happiness."

As long as GDP growth is defended with more passion than human happiness or the preservation of livelihoods, prosperity growth, which should not even be called that, will not come to an end, and neither will the climate crisis. It is therefore time for a paradigm shift. Time to clean up the capitalist paradigm of prosperity, which has no future on a finite planet.

Not only critical economics, political science, and environmental science are needed, but also media makers with their influence on public opinion.

VOLUNTARY SELF-DEPRIVILEGING

ALEX I felt a little weak in the knees when I was invited to the podium. Cameras, spotlights, upholstered chairs. Many people in shirts and suits were sitting in the audience. There was an air of renewal and celebration. It was April 2014, the year before the UN's Sustainable Development Goals were to be adopted. Everyone was to have a say in the so-called Future Charter, which redefined the guidelines of German development policy.

How did I end up on this podium? After refusing military service in 2008, wearing sandals and a peace dove on my shirt, I traveled to Peru to volunteer at a school on the outskirts of the main city of Lima. I spent twelve months doing theater with the upper grades, offering sports activities, and helping teach English to the first grades.

Now, a few years later in Berlin, I was invited as a former volunteer to discuss the Charter for the Future. It was supposed to be about the big questions, about a sustainable world and the steps toward it. With me on the podium were Development Minister Gerd Müller, the Co-President of the Club of Rome, Ernst Ulrich von Weizsäcker, and former Environment Minister and sustainability researcher Klaus Töpfer. Two young people who were active with UNICEF and the campaign organization ONE and I were introduced by Dunja Hayali as co-moderators. We, the young people, were allowed to ask questions, and the three gentlemen were supposed to answer them.

When I asked von Weizsäcker about the "limits to growth" that a study commissioned by the Club of Rome had identified in 1972, he mumbled something about "decoupling" and "factor five." These are the buzzwords of a theory according to which a decoupling of resource consumption and economic growth is possible. In other words, we can continue with GDP growth as before if we increase our energy efficiency by a factor of five. I was puzzled: von Weizsäcker, a mastermind of the energy transition and advocate of consistent environmental policy, outing himself as a disciple of technological and market-based solutions?

I was also shocked by the solemn confidence with which the panel spoke about the challenges of our future. All three, von Weizsäcker (SPD), Töpfer (CDU), and Müller (Christian Social Union, CSU) had stood out in their parties as progressive pioneers. Töpfer in particular remains an inspiring representative of conservative sustainability policy. And yet their argumentation remained trapped in the path dependencies of the status quo, in the belief in the old prosperity paradigm.

It was only on the way home that I realized I had sat on that podium with the wrong expectations. I had naively believed that the big issues were really being discussed there. Now I was shocked and disappointed at the same time. It was the beginning of a change in myself: The gap between the magnitude of the challenges ahead of us and the lack of political willingness to leave the path of least resistance made me more radical in the true sense of the word: "radix" in Latin means root, origin.

And one root of our misery undoubtedly lies in an economic system that can only continue to exist through constant growth. No tree grows to infinity, the space of nature is finite. So why should a system based on the exploitation of nature be able to grow indefinitely? I couldn't wrap my head around why the political answer to the problems of this growth model should be even more growth.

How can this change succeed? Luise Tremel, a scholar who studies transformations, points to the abolition of slavery as an example. Of course, the case could not simply be equated with today, but there are significant parallels, according to Tremel. Then as now, the economic system was based on exploitation: in the days of slavery on the direct exploitation of people, today on the exploitation of the natural resources of present and future generations. Just as society today is dependent on fossil infrastructures and energy sources, societies were dependent on the labor of the enslaved.

Because prosperity was based on this system, all people (except the enslaved themselves) participated in one way or another in the exploitation through slavery, as Tremel describes: "It was the business basis for agricultural production, transportation companies of all kinds, shipping companies, large and small commercial enterprises, insurance industries, banks, manufacturing industries of all stripes. Slavery therefore created and secured jobs in all of these industries. Tax revenues filled state coffers, private households derived comfort from domestic enslaved servants, and the great masses of people in these societies were able to afford sugar, cotton products, coffee, etc., in the first place because the actual costs of these products were externalized—outsourced to the enslaved."[15]

The parallels to the status quo are unmistakable. But as we know from our school lessons, slavery was not abolished because it became economically unviable. It was abolished because it was wrong and morally reprehensible. When the British Parliament decided to get out of the slave trade, the British were world leaders in the business. The majority of MPs did not emphasize that while the slave trade was a problem, going it alone nationally was no solution to such a global challenge. They did not point out that they were at a competitive disadvantage. Nor did they argue that other countries treated their enslaved populations much worse. They simply banned the slave trade, even though this was accompanied by economic disadvantages.

What makes this example a model for today's challenges is the fact that the enslaved could not free themselves from this exploitative relationship on their own. Except for a few examples such as the Haitian Revolution from 1791, in which enslaved people were successful in freeing themselves, their uprisings, escape attempts, and acts of sabotage could not bring down the system of coercion and violence. This changed only when people who profited from the system became active in overcoming this exploitation.

Without this "voluntary self-deprivation," as Tremel calls it, even today "a good exit from the transformation is hardly conceivable."[16] Since we must assume that those who suffer from our imperial way of life will not force us to change our path, the voluntary renunciation of privileges is an important step on the way to a just and sustainable society.

The word "voluntary" can quickly be misunderstood in this case: For unlike in the case of voluntary self-commitments of companies, whose systemic impact is minimal, we as a whole society must opt out of unsustainable practices. The initiatives of voluntary pioneers are important—even the legal abolition of slavery was preceded by boycotts, awareness campaigns, petitions, and protests—but we must all stop. The necessary self-deprivation is voluntary because no one can force us to do it except we ourselves. It is therefore a matter of consciously striving for less: less quantitative growth, less resource consumption, fewer emissions, less exploitation.

Ideally, the prosperity paradigm would combine the best of both worlds: high social standards, but low greenhouse gases and environmental pollution. Global relationships without global dependencies. Growth, yes, but only growth that increases equality and well-being. In other words, qualitative growth: in freedom, satisfaction, health, and independence.

DOUGHNUT FOR FUTURE

Such a paradigm shift requires a new way of economic thinking. It is, after all, our way of economizing that is a big part of the problem. One who has an answer to this is Oxford economist Kate Raworth. She answers the question about the future of the economy with a doughnut. And that's not a joke, but a brilliant approach. Imagine the shape of the doughnut—a small ring inside a large one: The doughnut defines a safe and equitable space that we should create for humanity with our economies. The outer ring is an ecological ceiling defined by planetary boundaries. The inner ring is the social foundation of well-being. Within the hole in the middle lie "critical human deprivations such as hunger and illiteracy."[17] The inner ring is a new way of thinking about economics.

A new way of thinking about economics in the twenty-first century would, according to Raworth, look something like this: Instead of increasing GDP, which sees economic growth as a measure of progress, the goal of our economic activity should be to meet "the human rights of every person with the means of our life-giving planet."[18] Moreover, the economy must be re-embedded in society and nature in order to overcome the "neo-liberal narrative of the efficiency of the market, the incompetence of the state, the confinement of the household to domestic life, and the tragedy of the commons."[19] The one-sided image of the *homo oeconomicus*, i.e., the image of a rational, utility-maximizing being with stable preferences, must also be overcome, according to Raworth. Because we humans are social beings who change our values and are interdependent with the living world, there is great potential, she says, "to nurture and develop human nature in ways that give far greater chance of getting into the Doughnut's safe and just space."[20]

Raworth also urges us to develop a method of systems thinking that leaves behind the mechanistic notion of an equilibrium of supply and demand. The economy is a dynamic process, complex and influenced in its evolution by social and ecological events. We should also dare to discuss distributive justice again. For Raworth, economic inequality is not an economic necessity, but rather "a design failure."[21] If we want to further close the gap between rich and poor, we should not only consider income, but also the wealth created by land, companies, technology, knowledge, and money creation.

While it is still widespread in economics today to see growth as a prerequisite for environmental protection, according to Raworth, in the twenty-first century we need a regenerative orientation of production and distribution that understands the economy as made of cycles and integrates "humans as full participants in Earth's cyclical processes of life."[22]

Because infinite growth is not possible in a finite world, it is pointless to cling to the growth of GDP and economies. Raworth therefore formulates our task as follows: "Today we have economies that need growth, whether or not they make us thrive: what we need are economies that make us thrive, whether or not they grow."[23]

THE "GOOD LIFE" AS A CONSTITUTIONAL GOAL?

The concept of *Buen Vivir*, or the good life, which found its way into the constitutions of Ecuador and Bolivia a few years ago, shows what a first step might look like. The approach is based on Indigenous traditions and values from the Andean region and has made the rounds as an alternative to the growth-based prosperity paradigm of the Global North.

In the Ecuadorian constitution, "Sumak Kawsay," as the concept is called in Quechua, was defined as a central constitutional goal in the course of a major reform. This not only includes the right to food, health, education, and water, but also a new understanding of development, as stated in Article 275: "*Buen Vivir* requires that persons, communities, peoples, and nations are truly in possession of their rights and exercise their responsibilities in the context of interculturality, respect for their diversity, and harmonious coexistence with nature."[24]

Thus, the driving forces of political action should no longer be accumulation and growth, but a state of harmonious equilibrium with nature. The concept is a conscious departure from the "higher," "faster," "further" of Western industrialized countries, which is expressed in the dogma of constant GDP growth.

Both Ecuador and Bolivia gave nature the status of a legal subject by anchoring *Buen Vivir* in their constitutions, which thus recognize nature for what it is: a living organism that is not to be protected for people, but for its own sake.

Another example can be found in New Zealand: In 2017, the government of New Zealand passed a law granting the Whanganui River on the country's northern island its own rights.[25] In doing so, it acceded to the request of the Maori, who had been fighting for 140 years to have the river legally recognized as a living entity. Since the river cannot represent itself in court, there have since been two official representatives, provided respectively by the New Zealand government and the Whanganui Iwi, who derive their name and way of life from the living Whanganui River.

It is examples like these that show how great the diversity of societal objectives can be when one dares to free oneself from the dependency of national economic growth. And they also show how crucial it is to think of the necessary changes in terms of constitutional reforms as well: An inclusion of "climate protection" in the German Constitution, as recently demanded by Markus Söder,[26] or of "sustainability," as proposed by the New Social Market Economy Initiative,[27] will, however, remain largely ineffective if no legal means are made available to sue for these goals if necessary. The widespread disregard for Article 20a, as we noted in chapter 5, shows how important concrete and binding targets are to achieve a change in the prosperity paradigm.

Numerous concrete proposals can be found in the Future Policy Award, which is bestowed every year by the World Future Council. Also initiated by Jakob von Uexküll, the award honors exemplary legislative initiatives that contribute, for example, to the protection of forests, oceans, and biodiversity. Laws on disarmament, ending violence against women and girls, and combating progressive desertification are also among them.[28] There is no lack of models, in theory and in practice, when it comes to reforming and revolutionizing paradigms of prosperity and economic practices. There is only a lack of willingness to embrace change.

Among White settlers the abolition of slavery was met with considerable resistance, not only before the decisions that led to abolition, but also afterward. The promised improvement took place only slowly and was hardly visible on the personal level. Hence when the cuts in their own privileges became noticeable, solidarity with the enslaved collapsed within a few years. A "social majority for abolition" became a "majority of people who find this change outrageous: too strict, too expensive, for a purpose now recognized as unworthy," writes Tremel. This resulted in outrage, resistance, and protests against the reform. As societal support fell away, so did political support for the enslaved, so democratic governments implemented the new order without much regard for those whom the regulation was intended to protect. This is because, according to Tremel, it was now no longer about the enslaved, but "about the lost prospects of prosperity for those who had benefited from the system."[29]

We observe something similar today: While the big profiteers of the current economic system are much more likely to be compensated for their inevitable losses through political pressure, as was observed with the nuclear phase-out and the outcome of the coal commission in Germany, for ordinary citizens changes often take place without such compensation. Since the reason for ending exploitative and destructive economic dynamics is at first glance only idealistic, but the reduction of privileges is immediately felt materially, we must understand which voices are not being heard, and develop strategies for cushioning the losses for those affected.

FOR A GREEN NEW DEAL

Does all this sound utopian? Not necessarily. The call for a "Green New Deal" in the United States, as put forward by Congresswoman Alexandria Ocasio-Cortez, among others, shows what concrete forms abstract utopias can take. If an economy is to become sustainable, Ocasio-Cortez's idea goes, it first needs an investment program. Investments should be made in sectors that generate a high level of social and individual satisfaction and are as low emission as possible. These include education, social services, health, and nature. The aim is to create low-carbon jobs and living spaces where people and environment are in harmony. She calls for measures to promote innovations toward a transition away from fossil capitalism and reports of coal workers planting mangroves after a retraining program. That is how the transition to a low-carbon society can succeed. Is this the future? At least these are the first steps.

Everything is possible. It all starts the moment we say goodbye to the blunt slogan that "climate protection endangers prosperity." It begins when we sincerely ask ourselves what real well-being actually is and realize that there can be no good life on a destroyed planet. That's when we roll up our sleeves.

9 THE CLIMATE CRISIS IS A CRISIS OF JUSTICE

ALEX In Berlin's Invalidenpark, where the climate strikes take place every Friday, I stood next to a group of mothers who were attending for the first time. Together with hundreds of others, I chanted, "What do we want? Climate justice! When do we want it? Now!" The women, who had remained silent, looked around questioningly. Climate justice? What is that supposed to be? I wondered how I could explain it to them in three sentences. Where do you start? How do you explain that the climate crisis is not only a threat, but also an injustice?

Not everyone has contributed equally to the climate crisis. Not everyone is equally affected by it. Therefore, the climate crisis is also a crisis of justice, both in terms of its causes and its consequences. That is why dealing with the crisis is always a moral question as well. It challenges us not only as citizens, parents, or consumers, but above all as moral beings. And because causes and consequences are unequally distributed, the solutions must be just.

The climate crisis is also unjust because it exacerbates existing injustices—not only between the different generations, but also between the Global North and the South, between the rich and the poor, between people of different skin colors, between urban and rural areas, and between the sexes. People with disabilities will be particularly hard hit by the consequences, and the elderly will also suffer more than others from extreme weather fluctuations.

This is the reason why intersectionality, i.e., the intersection of different experiences of discrimination in one person, plays such a big role in all of this; the crisis creates new inequalities. Young people are burdened with the unpaid ecological bills of the elderly; but the crisis above all reinforces existing inequalities. Poorer people, for example, are less likely to be able to afford to move when natural disasters strike, and their property is less likely to be insured, so the climate crisis makes them even poorer. Moreover, the various discriminations of different social groups interact with each other: For example, if you are young *and* comparatively poor, you experience a different form of "climate discrimination" than if you are young *or* poor.

Fundamentally, the dynamics of this injustice, which we call climate discrimination here, are mundane: The less resilient you are—economically, socially, or culturally—the more the climate crisis hits you compared to others.

The example of Hurricane Katrina, which in 2004 destroyed four-fifths of the US city of New Orleans, shows how dramatically poverty can affect the consequences of extreme weather events. Entire neighborhoods were washed away, and the damage was greatest where the city's Black population lived. Because Black people in particular lived in poverty, they were the ones who had no cars to flee the waters, lost their homes, and sought emergency shelter in large numbers.[1] The White population that fled the waters was transported an average of 193 miles away, the Black population 349 miles. Many of those who made it back to their homes afterward found boarded-up doors and police forces refusing them entry. The city demolished the vacant public housing, nearly one hundred percent of which had been occupied by Black people, to make room for new construction. Due to the high insurance sums that people have to pay today after the hurricane—the damage is estimated at $125 billion—rents have risen by a third, and homelessness has doubled. The city's Black middle class has since shrunk dramatically, and the Black share of the population as a whole has fallen from 64 to 59 percent.[2]

What the Katrina example also shows is that the consequences of global warming reinforce racist structures in society. Because in many places people of color are forced into or kept in precarious conditions due to structural and institutional discrimination, it is to be expected that they will be disproportionately affected by climatic changes. Until these structures change, the climate crisis will also be a racist crisis.

Wherever we look, the climate crisis threatens to reinforce existing inequalities. Respect for human rights and much of what progressive movements have fought for in recent decades will come under increased pressure if we do not mitigate the consequences of climate change ecologically and socially early on. Recently, Philip Alston, the United Nations special rapporteur on extreme poverty and human rights, expressed fears in a report that human rights will not survive global warming and that we are sliding into an era of increasing climate apartheid. The progress of the last fifty years in development, poverty eradication, and health, he worried, could be lost in the climatic changes. What Alston describes is a grim scenario: a world in which the rich pay to be spared heat, hunger, and conflict, while the poor are exposed to them unprotected.[3]

THE PRICE OF FOSSIL PROSPERITY

LUISA Through the large office windows, you can see the postcard-blue sky and the flags of the Bundestag moving in a lazy breeze. Black, red, gold. It's early July 2019, and Berlin is melting. My pants stick to the leather cover of the visitor's chair. I'm wearing sandals, despite the occasion. Even Christian Lindner has sweat on his forehead, although he otherwise looks like he's jumped out of an FDP (Freie Demokratische Partei, a conservative market-liberal party) campaign poster. We are sitting here to talk about the climate. Across from us sits his five-person team, the "Team Lindner." Somewhat disproportionate, I think to myself, for a simple podcast recording, because this video format is rarely clicked on more than 5,000 times on YouTube. I don't yet know that our conversation will find 70,000 viewers within a month.

Mr. Lindner had invited me to his podcast after I had accused him on Twitter of defaming those who focus on the essentials with his eternal buzzword bashing (against bans, for freedom, and so on). I thought that was a pretty savvy move on his part. You have to be able and willing to do that: giving someone with whom you have such obvious differences space in your own podcast. I had spent an afternoon working my way through a stack of reports, party platforms, and interviews in order to prepare.

While we were busy shaking hands, asking for coffee, and getting wired up with microphones, I wondered who would end up accusing me of letting myself be used for a PR stunt by the FDP. Possibly quite a few people. I therefore took it upon myself to make the conversation worthwhile. I wanted to discuss things as constructively as possible, and resist the temptation to take an overly polemical approach to Lindner and his party.

Such conversations are a tightrope act. Up to that point Christian Lindner had shown minimal interest in the climate crisis. Since the signing of the

Paris Agreement, he had delivered four party conference speeches. These speeches, each well over twenty pages on paper, were Lindner's opportunity to lay out the FDP's program and indicate their future priorities. Between 2016 and 2018, when Greta Thunberg was still an unknown, Lindner had mentioned the word "climate" only once. That was in 2018 when he had said, "Yes to a common European energy and climate policy."[4]

When people who present themselves as experts on the future and talk about new trends or, in Lindner's parlance, "updates" that the country needs—when such people exclude the crucial cornerstone of future-proof policy of all things, I find that disconcerting. As a rule, I suspect that these people are not that familiar with the future after all. That's not meant to sound condescending. If Lindner had really been familiar with the state of science, he would probably have talked about the climate even before it was possible to win elections with it. Or would he have? At least I would have hoped so.

When the climate became a hot topic in 2019, Lindner also spoke about it at the FDP party convention. He denounced the federal government's "planned economy" in climate policy, talked about a "highly energetic" debate and "hyper-moral" decisions. He compared the issue to the "refugee crisis" of 2015, when the decision was made to open the German borders.

So now Lindner was sitting next to me claiming that he and his FDP were of course committed to the Paris Agreement. Did he actually know that trying to limit global warming to 1.5 degrees meant that Germany would have to achieve zero emissions between 2030 and 2040, which was obviously far sooner than the German government had stated in its own targets? That a commitment to Paris meant there was also a commitment to the aspect of fairness? From which it follows that Germany would also support other countries financially in their climate protection measures in the long term? That this agreement meant that one could not blindly rely on the fact that emissions would fall solely through "free will" and "market mechanisms"?

On the afternoon in question, I experienced a conspicuously reticent Lindner; one who frequently asked questions and only dispensed his typical liberal interjections in passing. We talked for a long time about tax reforms and market mechanisms, and I explained a bit about the Paris Agreement. Time dragged on, the sun shone through the windows, and we sank deeper and deeper into our chairs. Even the flags on the Bundestag seemed demotivated and were only drooping lazily. That's when it finally got interesting.

"What would be really sustainable and fair is to start saying now, 'Okay, we need a carbon price that in some way bears the whole cost.'" That's what I say about the proposal to put a price of 180 euros on a ton of CO_2 because, according to the German Federal Environment Agency, that's the amount of the costs that result from it. Lindner's spontaneous reaction: "Well, 180 euros of course includes the tons that our parents, grandparents, and great-grandparents helped emit. We've been partying here for a hundred and fifty years, you said, and we've emitted CO_2. And if we now say that the German CO_2 avoidance price should be 180 euros, then that's a price that, let's say, includes past CO_2 emissions—because we could emit more CO_2 now if we hadn't emitted so much twenty or thirty years ago." Now Mr. Lindner starts rubbing his eyes. But that's the way it is: We have to pay the price for what generations before us have screwed up. We're paying the price not only for our prosperity, but also for earlier fossil fuel prosperity. True: That is unfair. But it will become even more unjust and unfair as long as we do not face up to this fact. That is the core of the climate crisis. The core of what millions of young people are now demanding Friday after Friday.

The injustices brought about by the climate crisis, in all their diversity, are impossible to capture in a nutshell. Moreover, in a privileged position like ours, it would be presumptuous to claim to be able to understand the various ways in which people around the world are being discriminated against by the climate crisis. What we want to do here instead is to highlight some of the dimensions of this injustice—in the hope of sensitizing people and widening their view of this moral challenge that must be at the core of all climate change efforts.

INTERGENERATIONAL JUSTICE

Intergenerational injustice is reflected in the concept of sustainability, which has been talked about for several decades. Its central idea was articulated in the seminal Brundtland Report of 1987: "Sustainable development is development that meets the needs of the present without compromising the ability of future generations to meet their own needs."[5]

The climate crisis testifies to the failure of this sustainable development: We, the young of today, have to pay for what the generations before us have done. And our descendants have to live on the planet that we are bringing out of balance. That's why we talk about stealing the future (chapter 2), that's why the climate crisis is also a crisis of responsibility (chapter 5). The state objective in Article 20a anchors the idea of intergenerational justice in the German Constitution, but it is still being sacrificed to short-term economic interests and the arbitration of conflicts of interest.

The climate crisis exceeds the foreseeable time frame of legislative periods; since the consequences of rising emissions will only become apparent in the near and distant future, and climate protection brings comparatively few benefits in the short term, the temptation to postpone investments in climate protection is great. But the effects of the crisis are moving ever-closer to the present. This is precisely why we young people are the first generation to demand climate protection out of direct self-interest: We are the first to experience any impact of climate protection, or in the worst case, the catastrophes that will result from a failure to protect the climate.

CARBON JUSTICE

In November 2017, a group of people visited the village of Morschenich on the edge of the mining area of the Hambach open-pit lignite coal mine in the Rhineland. As the visitors turned their gaze to the devastated landscape, they began to cry. The scene was captured by camera crews, and shortly afterward videos showing this memorable moment went around the world. These were unusual visitors. They had come from far away, from several island nations in the Pacific; they joined together as Pacific Climate Warriors because their homeland is threatened. With every millimeter that the sea level continues to rise, they lose a piece of their home. In the Rhineland, ten thousand miles from their home islands, they found one of the causes of the catastrophe. "German lignite mining operations are among the largest in the world. If we don't close them down," said Zane Sikulu from Tonga, "the Pacific Islanders won't stand a chance."[6]

Ten years earlier, in 2007, Angela Merkel was still considered a climate chancellor. At that time, at a climate conference in Potsdam, she called for clear reduction targets, clear responsibilities, and the long-term alignment of per capita emissions. Merkel warned of dramatic losses in prosperity and pleaded for "carbon justice" between rich and poor countries. Merkel did not create the term carbon justice. Wangari Maathai did. Maathai was a Kenyan activist and founder of the Green Belt Movement in Kenya. The movement has planted more than 51 million trees since its beginnings in the 1970s, trained thirty thousand women in forestry and beekeeping, and created thousands of jobs.

What Maathai meant when she spoke of carbon justice is recognition of the circumstances under which nations are moving toward a carbon-neutral economy. After all, many countries, particularly in Africa, are feeling the effects of global warming more than others, even though they have contributed very little to it. Of the victims of extreme weather events since 1980, 80 percent have been in Asia, Africa, and Latin America. This is not only because more people live there, but also because many people there lack the resources to cope with extreme weather or adapt to new conditions.[7] This trend is set to worsen. According to Philip Alston, so-called developing countries will bear three-quarters of the costs, even though the poorer half of the world's population is responsible for only ten percent of CO_2 emissions.[8] Thus, those who have contributed the least so far are particularly vulnerable to global environmental change.

Water scarcity, for example, will first worsen for people in the Sahel, southern Africa, Central America, and the Mediterranean; agriculture in subtropical and tropical drylands will suffer more from seasonal shifts and declines in rainfall. The countries of Southeast Asia will have to cope with irregularities in the monsoon, and floods will mainly affect the large delta regions such as Bangladesh and India. Rising sea levels pose a particular threat to island states in the Pacific or cities such as Somalia's capital Mogadishu, which lie at sea level.[9] Diseases, insects, or pests that occur more frequently due to the climate crisis or spread to places where they were previously unknown will also become a threat, especially in regions where few economic resources are available to combat them or take preventive measures.

Against this background, the researchers at the Potsdam Institute for Climate Impact Research speak of a historic responsibility of the industrialized countries. The reduction pathways presented by the Intergovernmental Panel on Climate Change for the 1.5-degree target—so-called "Shared

Socioeconomic Pathways" (SSPs)—therefore provide for a distribution of tasks: Since the social and economic situation of countries must be taken into account in the climate targets, Western industrialized countries must aim for an immediate and drastic reduction in emissions, while sustainable development in countries of the Global South must first aim to improve the living conditions of their populations in the most climate-friendly way possible. According to the IPCC, reducing inequality between countries could actually reduce the emissions intensity of global economic growth.[10]

In other words, because the industrialized countries have been emitting greenhouse gases for more than two hundred years as a result of strengthening their privileges at the expense of others, they now have a historic responsibility to avert the catastrophic consequences of this way of life. On the other hand, those countries that have been forced into economic and political dependency for centuries by the imperial way of life of the industrialized countries should be given the chance to dedicate themselves to the fight against extreme poverty, inequality, and other tasks through more moderate emissions targets.

However, historical responsibility is often obscured by referring to current emissions per country. It is the emissions per capita that show the differences between rich and poor countries. The per capita consumption of CO_2 was around two tons in India in 2016, while each German consumed an average of nine tons of CO_2 in the same year.[11] Of course, the current emissions of countries are an important factor in the political debate. For as long as the United Nations cannot legislate and punish violators, nation-states are the actors who must enforce emissions reductions. But from a justice perspective, we must not lose sight of cumulative emissions and emissions per capita.

If we take human rights seriously, and if we see all people as equal no matter where they are born, we cannot morally justify that a few live beyond their means and thereby deprive others of the opportunity to live a dignified life. If we must not exceed an annual CO_2 budget of 1.61 tons per person by the end of the century in order to comply with even the 2-degree target,[12] countries such as Australia (approximately 17 tons per person in 2017), the US (approximately 16), Canada (approximately 16), and also Germany have a greater responsibility to reduce greenhouse gases. In theory, this also applies to countries such as Estonia (approximately 15), Kazakhstan (approximately 16), or Qatar (approximately 50), but the impact of their emissions policies at the international level is not comparable.[13] And as climate researcher Johan Rockström repeatedly emphasizes, "If Germany can do it, others can do it, too. If Germany fails, the odds are against everyone."[14]

A SEXIST CRISIS

"The climate crisis is a man-made problem and must have a feminist solution," Mary Robinson, former UN high commissioner for human rights and the first female president of Ireland once said.[15] Her statement became world famous.

For decades now, Mary Robinson's work has put an issue on the agenda that had long been neglected—namely, the role of women in both the climate crisis and climate action.

The fact that women are fundamentally structurally discriminated against is unjust enough, and demands change, quite apart from the climate crisis. The climate crisis, however, reinforces many of these inequities—making the climate crisis a sexist crisis.

The baseline is not unfamiliar—women earn on average 23 percent less than men worldwide, do between two and eight times as much unpaid care and childrearing work, and have longer workdays—which means they work a combined four years longer in their lifetime than men, according to a 2017 Oxfam study.[16] That year, according to a survey by the consulting firm Deloitte, just 15 percent of global board seats were held by women.[17] Figures like these, which could fill many pages, naturally vary from region to region. Together, however, they paint a clear picture: Globally, women are more often affected by poverty and socioeconomic insecurity and are less likely to be in positions of political or economic power. All these inequities are exacerbated the poorer the women are and the less rich the country in which they live.

As a consequence, vulnerability to the climate crisis is higher. An accumulation of many structural inequities interacts with the climate crisis to turn an already poor starting point into a sexist climate catastrophe.

For example, when droughts cause resources to become scarce and prices to rise, or extreme weather damages property or one's subsistence economy, it is those with poorer financial cushioning who suffer first—and these are more often women than men. When the climate crisis makes people sick[18] and poor, women are more often affected. In the Global South, the majority of agricultural work is done by women, so weather variability and shifts in vegetation cycles hit them harder than men. When the climate crisis causes extreme weather and natural disasters that kill people, it is more often women who die, and more often women who die younger than men their age. If the climate crisis and the resulting conflicts over lack of resources force people to flee, 80 percent of those affected are women. The Paris Agreement also notes that women are disproportionately affected by the climate crisis and that their empowerment is critical to upholding the agreement. Still, the role of gender in the climate crisis is a grossly underestimated dimension.[19]

In short, women worldwide are among the big losers in the climate crisis and hardly anyone is talking about it. And who, structurally, are the winners? Men. More precisely: older white men from the Global North. For it is still they who are disproportionately to be found on the boards of energy companies, in the political establishment, and at the levers of the financial industry, generating returns, revenues, and power through the burning of fossil fuels. The big faces of this are the Trumps and Bolsonaros of this world, but they also exist on a small scale, in every community, in every company. In this context, "man-made" takes on a whole new meaning.

LUISA A large theater had invited me to discuss a play after the performance. It was an exciting setting, after the actors had left the stage, to sit where just a few minutes before great stories had been told. The play had dealt a lot with sexism, and we on stage were talking about it, too. I talked about how women working on small farms in Africa suffer more than male farmers from climate change impacts like water shortages, because it's traditionally their responsibility to fetch the water. I talked about how on average more girls get sick or even die in the increasingly frequent extreme weather. I talked about how depressing it is to realize that women and girls are on the front lines when the climate crisis hits the world.

Afterward, an older woman approached me, purposefully interrupted my conversation, and said firmly: "We don't even have to look all the way to Africa to see this blatant injustice: What's stopping us from doing anything? The jobs! And whose? Those of the men who work in coal and automobiles. If women were employed there, these jobs would have been gone long ago!"

She raised her index finger, and I inconspicuously moved back a step: "For decades, men have fueled climate change because it brought them money and jobs—and they didn't think it was wild that nothing was being done. That's never going to change unless you get women to do it."

It was the idea of "getting women to do it" in particular that moved me for a long time after the conversation. Yes, we should get the women to do it; a future can only be justified if the discriminated half of the world's population helps to shape it. But where were the women who were supposed to participate? I thought of the many people who had called me in the months before to tell me that they also had ideas about tackling the climate crisis. They would often say, "I want to save the climate," and present me with their plan for climate action. I had had well over a hundred conversations like that.

And it struck me that I could count the conversations with women on one hand.

Mary Robinson hit the nail on the head with her statement: The answer to a crisis that is sexist in its origins and consequences must be feminist. This does not mean excluding men, but coming together with them on an equal footing. Up to now, women have been more often excluded from political participation and decision-making processes, although their participation often leads to more effective cooperation between diverse groups and to sustainable results.[20] This must change. It is both about the fact that in the midst of one of the greatest crises, the world simply cannot afford to do without the creative skills, energy, and ingenuity of half the world's population. But it is also about the fact that, for example, long-term food security will not be steered in sustainable directions if women are not brought to the table. Climate justice therefore also means getting women out of the firing line of the climate crisis, engaging their perspectives and experiences, facilitating access to education, and empowering them politically, culturally, socially, and legally.

WHO IS BEING HELD ACCOUNTABLE?

When we call for climate justice, we should be vigilant about who is held accountable for actions to mitigate the climate crisis. When large-scale reforestation programs are started in the Global South or dams are planned, it is not uncommon for them to endanger the lands Indigenous or other vulnerable groups call home, as we know from the past. The Sardar Sarovar dam in India's Narmada Valley is one such case. Part of a mammoth project involving thirty large dams and over a hundred small ones, it became a symbol of unsustainable development in the 1980s. Begun in 1961, the construction of the dams displaced over a million people, mostly small farmers and Indigenous peoples, from their homes, flooded entire forests, and destroyed the habitats of rare species.[21] However, the people affected got organized and, despite multiple repressions and arrests, won a Supreme Court ruling in their favor, which led to the interruption of the dam construction project. Medha Patkar, who received the Right Livelihood Award for this in 1991 together with the Narmada Bachao Andolan movement, is still active today in the struggle for an inclusive, democratic, and environmentally friendly development of the region—and to this day the over-sixty-year-old is exposed to repression by the authorities. What this story shows is the need to think through the social consequences of the policies we undertake to combat the climate crisis. The urgency of projects that produce effective emissions savings should not lead to the exclusion of basic democratic principles or human rights in planning or implementation.

THE NEW SOCIAL QUESTION?

What is true on a global scale is also true for inequalities within individual societies. For the enormous transformations demanded by the climate crisis have the potential to exacerbate many existing injustices.

It starts with who has the opportunity to have a say in the design of climate change plans—rarely do the perspectives of the less privileged get heard. People with lower incomes also have fewer opportunities to adapt to the changes that are coming or have already occurred. If the measures for adapting to the consequences of the climate crisis—because it is no longer possible to completely prevent them—are not understood as a collective task of the municipalities, states, and the global community, then existing inequalities will become even more pronounced.

If the political decisions made to reduce emissions do not take who will be disadvantaged by them into account, a loss of trust in the government and the political system is hardly avoidable. The yellow vests in France are an example of the political earthquake that an unjust climate policy can trigger: Especially people in rural regions, living on low incomes and dependent on a car because of where they live, were affected by the fuel tax increase; for quite a few it posed an existential threat. A government that wants to protect the climate at the expense of society's marginalized endangers the foundations of the democratic polity. Anyone who talks about reduction targets must not forget about redistribution. This is also what we mean when we call for climate justice.

Effective taxes on climate-damaging materials and practices, such as the French tax increases on diesel and gasoline, are likely to be important building blocks for reducing emissions, but they should be paid primarily by those who contribute most to the climate crisis and who do not expect to suffer any existential losses as a result of the levies.

Consumption taxes on everyday goods always run the risk of perpetuating existing inequalities, as they affect those who have little money at their disposal—see the protests of the yellow vests. The repeated increase in the value-added tax in Germany is another example of this. After the black-and-yellow coalition[22] simply phased out the wealth tax in 1997, thus forgoing annual tax revenues equivalent to more than 4.5 billion euros,[23] the value-added tax was gradually increased from 10 to 19 percent. While the superrich in our society were relieved of tax burdens, the government increased tax revenues in such a way that it was primarily those who had to turn over every penny twice who noticed.

A wealth tax had existed since 1923, and it still exists today under Article 106 of the German Constitution. In the 1950s and 1960s, it contributed just under 2 percent of tax revenue; until 1996, the tax was 1 percent of wealth. According to a study by the German Institute for Economic Research, reintroducing the tax as it was levied until 1996 would generate annual revenues of ten to twenty billion euros, even with high exemptions.[24]

We are faced with the task of organizing a massive restructuring of our energy, mobility, and production infrastructures. This requires not only concepts, but also money. That has to come from somewhere. And it would be fair if especially those who profit most from the unsustainable status quo were to pay for it, too.

However, we also mention this because we must realize that we will hardly be able to achieve a just transition to a climate-neutral society by 2050 without asking early on how we will finance it. Wealth taxes or climate-related wealth levies are two answers that have hardly been discussed so far. In January 2022, over 100 millionaires called for a permanent wealth tax in an open letter to government and business leaders.[25]

Another example of the social dimension of actual challenges: The contribution of air traffic to global warming is estimated at 5 percent.[26] If flying becomes more expensive, whether through the introduction of a tax on jet fuel, an effective CO_2 price, or other measures, people with low incomes in particular will be able to fly less or not at all. While only 18 percent of the world's population has ever boarded an airplane and only 3 percent of the world's population flew in 2017, flying has now become a mass phenomenon in Germany. In the year 2018 a record was set, with over 80 million passengers taking off from German airports in the summer alone.[27] Globally, air traffic has increased by more than two-thirds since 2005. By 2050, the volume is expected to increase three- to sevenfold.[28] And despite successful campaigns like #flightshame, which is already having noticeable effects in air and rail travel in Sweden,[29] a reduction in emissions cannot be expected without clear regulations.

Kenyan Wangari Maathai showed how awareness of injustice can be used as a benchmark for responses to the climate crisis: Her Green Belt Movement both empowered women and local communities and contributed to food security, as well as raised environmental awareness and created job prospects.

10 EDUCATE YOURSELVES!

It is tempting to blame the inaction of many on the lack of knowledge about the climate crisis. It is tempting to believe that people would act if only they "knew better." One might think that the call to "Educate Yourselves!" is just that: That people just need to understand better what is happening to the climate—then they will act. But it is not that simple.

The majority of Germans are more or less aware of the climate crisis. Studies have repeatedly shown that there has long been an abstract awareness of the problem. According to the European Social Survey, 53 percent of Germans state that they are strongly or very strongly concerned about the climate crisis. This is higher than in almost all other European countries, with the exception of Switzerland. Among Germans, 45 percent are very or extremely concerned about the climate crisis; 71 percent say it is the greatest threat to society, ahead of ISIS terror, cyberattacks, or the North Korean nuclear program.[1]

Even business-related institutions such as the World Economic Forum now clearly state that the greatest risks for the global economy are the lax handling of the climate crisis and the resulting catastrophes.[2] In April 2019, BlackRock, the world's largest asset manager, also warned of the increasing risks to investments that are to be expected as a result of the climate crisis.[3] It was like a small revolution that a conservative institution from the financial sector was warning of climate damage in such clear terms. In addition, countless prominent economists, such as the British economist Nicholas Stern, are questioning the myth of infinite fossil fuel growth and warning the business world against irresponsibility.[4]

By no means are all people educated, and many of those who consider themselves well informed are not well enough informed. But the sober observation remains: Enough people have known about the crisis for long enough.

Nevertheless, almost nothing is happening at the political, economic, or even individual level, or rather, nothing that does justice to the scale of the threat.

While ecosystems are collapsing faster than they can be fully explored, and environmental catastrophes are endangering people on an ever-more-threatening scale, half-hearted coal compromises, new packaging guidelines, and e-scooter laws are celebrated as successes. Because there is nothing else to celebrate.

THE GAP BETWEEN KNOWLEDGE, PERCEPTION, AND ACTION

The great climate policy moments of the recent past were those when common targets were agreed upon. In 2015, the 1.5-degree target was agreed upon in the Paris Agreement; in 2016, the German climate target was set for 2030; in 2018, the European target for 2050, updated in 2021.[5] Setting targets is important. But meeting the agreed targets is the really ambitious, the really important part. And very few governments and states manage to do that.

Beyond the political arena, for example in the private sector, there is no sign of any development in Germany, Europe, or internationally that would lead to concrete action to reduce emissions on the basis of the abstract awareness of the problem. Although the smallest levers can be found in the private sphere, individuals can make a difference. One of the most feasible and at the same time most effective measures is to abstain from meat consumption or at least to reduce it drastically. However, while half the people in Germany recognize the climate crisis as a major problem, according to *Statista*, only about 6.1 percent of people in Germany were vegetarians in 2019.[6] While this was a higher percentage than the previous year, it was lower than in 2010.

One could fill many pages with examples of inadequate action being taken.

It starts with the fact that information about the state of the global climate is first checked to see whether it fits one's set of values. For many people, this is where it ends. For example, if I choose to believe that a combination of "technology" and "market mechanisms" will solve the problem, simply because that has been my answer to political problems for twenty years, then I will dismiss all further details about the crisis as "scaremongering"—even if scientists speak in all clarity about the fact that we need comprehensive transformation beyond known measures. If I decide for myself that I find it exaggerated to talk about the potential extinction of parts of humanity, or about the fact that millions of people will lose their livelihoods, then I am also prepared, in case of doubt, to question the messengers of this information—for example, the scientists who suggest just that based on solid reasoning, or people like Greta Thunberg.

But anyone who takes seriously what science tells us about the climate crisis faces a different problem: What to do with this information? How does one deal with information about a potentially life-threatening danger?

"Climate anxiety" is the name of the phenomenon that more and more people are experiencing. It leaves especially the well-informed and consequently rather depressed people in a kind of climate powerlessness. Young people in particular are increasingly confronted with it. Others are processing the impact of this information by storing it away as "too complex." This, in turn, can lead to the conclusion that nothing can be done about it anyway—and that there is consequently nothing we can do about it ourselves.

Probably the most disastrous conclusions to be drawn from information about the climate crisis are that the problem is too big to be solved, that one is not really impacted by it, or that others will take care of it.

Even those who proceed from understanding to taking concrete action often do not realize that their actions as a whole are far from adequate. They are well informed about the dangers of the climate crisis, see themselves as environmentally aware, and view their behavior as largely sustainable—even if it is not. This "mind-perception gap," i.e., the gap between ecological awareness, the corresponding behavior, and its perception, is the last bottleneck on the way to a climate-neutral way of life. Annett Entzian, who has done research on this gap, sums up this problem with the phrase: "For they do not do what they know."[7]

Thus, it's not mainly the communication of information that is missing. The fact that too little or nothing is being done also has to do with the fact that people are not informed about how the beginning of the end of the climate crisis could be initiated. There is a lack of knowledge about the actors who are fueling the climate crisis and a lack of self-awareness about one's own role in bringing about change.

When we write "Educate Yourselves!" we do not (only) mean that the scientific background of the climate crisis needs to be taken more seriously, but above all what must come next.

"What is happening right now?"—this question can be answered.

"How do we change this?"—that's where we start.

"What would be possible if we really wanted it?"—We must devote ourselves to answering this question. And no longer hypothetically.

What does the call to "Educate Yourselves!" mean in concrete terms? It starts with the realization: It's going to be long, it's going to be tough, and it's going to take all of us. Here are five first steps:

1. *Educate Yourselves about How to Educate Others*

We need you to become professionals! Write, speak, and educate others about the climate crisis. But take it seriously.

Yes, it is necessary to enlighten people about the climate crisis. But if you take climate science seriously, you should also take climate communication seriously, and invest in it. Because the most important findings will remain meaningless if they are not communicated in a sensible way. For research institutions and donors, this means investing consistently in professional communication of research results. Not in the form of links to a PDF document, but in the form of multimedia design, in real communication online and offline that is engaging, vivid, and clear. As a scientist, you may find it incomprehensible that people are not shocked when they learn that 95 percent of the moors in Germany have been drained. To most people, this means nothing, it does not touch them and does not make them understand the drastic nature of the situation, nor does it inspire them to act. Accordingly, understanding the fact that people accept, absorb, and comprehend some facts better than others must be the basis of our work. This also applies to media and influencers.

The media, in turn, must adapt their professionalism in terms of climate communication to the reality of the crisis. Two examples: Almost all newspapers have a sports section, but hardly any have a climate section. As a result, there are sports journalists who are trained to talk professionally about soccer, boxing, and tennis, but there is no corresponding model for climate journalists. Every day, the nearly ten million viewers of the daily news are informed by a professional about the day on the stock exchange, which for most feels like a parallel world. The state of our livelihoods, on the other hand, is only mentioned when a foreign correspondent is allowed to speak for a few minutes about a catastrophic flood.

If this is not addressed with a professional approach in newspapers, on television and radio, but also in NGOs, networks, and institutions with a wide reach, too little will change in the next thirty years. The knowledge about the climate crisis is available. Now it needs to be communicated in a relatable way. That starts with a language that can put what is happening right now into words. And because there has never been anything like it before, we need a new language for it. So please: Breathe life into the information.

2. Tell the Truth, the Whole Truth

After the big energy companies in the US began to systematically spread doubt about the climate crisis in the late 1980s, talk about the climate crisis experienced the full spectrum of truth distortions—to the point that two climate-change deniers have been elected presidents of two of the most relevant countries in the fight for a healthy planet: Donald Trump and Jair Bolsonaro.

As a consequence, communicating the truth has taken on a new meaning. Because it's not just about reporting. It's about asserting oneself, being louder, clearer, and more articulate in order to penetrate through the persistent buzz of those who package lobby interests as climate facts and spread falsehoods.

The point is twofold: first, to factually address climate science and its consequences for people. And secondly, it is about exposing those who play down the climate crisis and those who drive it forward, those in politics, business, the media, and society who are responsible for delaying and obstructing climate protection.

It is no coincidence that the climate crisis has long been ignored, that climate targets are set too low to comply with the Paris Agreement, that fossil fuel infrastructure continues to be financed and built, that the German energy transition has stalled, and that climate protection continues to be postponed until it is unaffordable. The list could be continued. Behind it all are decisions favoring the short-term interests of a few against the long-term interests of the many.

It should not be a secret who is involved in these decisions. All those, be they political, economic, or financial entities, should be held accountable. Therefore, we need to start talking about them. Or to put it in the words of Jean-Paul Sartre: "*Connaître l'ennemi! Combattre l'ennemi!*" Know your enemy, fight your enemy.

In order to have honest debates we must know who is pulling which strings and how bad the situation really is. The big question of how to shape the next two, fifteen, thirty years demands a thorough debate with a great diversity of participants representing various perspectives. But these debates need a common factual basis, which at present is continuously challenged. At the same time, the actors behind the transformation are constantly under attack. Educate (yourselves) about the situation.

3. *Educate (Yourselves) about the Beginning of the End*

Imagine the climate crisis could be overcome, but no one would believe it possible. We would fail. That is exactly what is happening right now. Those who do explain the situation usually talk about the drastic nature of the crisis. What is needed is information about how to move forward, a training of the imagination, an understanding that learning about the climate crisis is just as important as learning about the way out, about the beginning of the end.

We could tell inspiring stories at this point, of role models and landmark laws that show how things can be done differently. The story of the Spanish minister for ecological change, for example, who negotiated and implemented the immediate closure of the ten remaining coal mines in 2018 shortly after taking office—with an agreement that enabled the workers to maintain their standard of living, retrain, and obtain new perspectives. These examples are important because they make visible how many positive things are already happening.

The climate scientist Friederike Otto and her research team have developed the field of "attribution science." Through complex calculations, the team can calculate whether, and if so, how strongly an extreme weather event is human-made, i.e., influenced by the climate crisis. In the best case they can do so within a few days. What may seem banal is actually revolutionary. Imagine: We can now know how much the climate crisis is responsible for every flood, drought, or storm.

This not only provides clarity about how dramatically the climate crisis is already affecting us and how great the danger really is. For the first time this scientific discipline also makes it possible to account for damages caused by extreme weather events. Climate litigation, i.e., the role of courts in climate protection, will be instrumental in bringing the climate crisis to an end.

So far, governments, companies, and investors have been able to fuel the climate crisis and delay climate targets more or less unchallenged. Those affected in the Global South, or we, the young and future generations, who already feel or will feel direct impacts, usually have insufficient means to claim our rights in court. With more and more concrete scientific findings about the severity of the crisis, with ambitions and creativity in jurisprudence through increasing application of the findings of "attribution science," and with a growing understanding of which actors are driving the climate crisis and in what form, the starting position is already changing bit by bit. Already there are over a thousand lawsuits underway in twenty-eight countries.

Even independent of "attribution science," political and economic entities are being sued more and more frequently. Take, for example, the energy company RWE, which is responsible for 0.47 percent of the emissions in the atmosphere (for comparison: The corporations Exxon Mobil and Chevron are in the lead with 3 percent each).[8] That the attribution of responsibility can have legal consequences is shown in the case of Saúl Luciano Lliuya. The Peruvian farmer sued RWE. His village is threatened by a dam that could burst due to continuing glacier melt. He is now fighting in court in Germany, supported by the NGO Germanwatch, for RWE to pay 0.47 percent of the costs of rebuilding the dam.[9]

If climate protection were to be included in Germany's constitution, which is long overdue, there would be an opportunity to sue for the right to an intact climate, regardless of who is in government.

This could also be applied to sea-level rise and pave the way for a new understanding of the responsibility for climate damage. Who should pay for climate damage, i.e., for what is called "loss and damage" in the context of climate policy? Costs will rise to astronomical levels as extreme weather becomes more frequent—consider, for example, Hurricane Harvey, which swept across the southern United States in 2017 and is estimated to have cost $190 billion in damages, making it the costliest storm in US history.[10]

Data will also have a prominent role to play in the task of addressing the beginning of the end. Not only data on the impacts and implications of the climate crisis, which are critical in the context of adaptation measures, but also data about mitigation. While technological developments are often celebrated as the answer to the climate crisis, increasingly precise data sets on CO_2 sources show that the most favorable and rapid measures often lie elsewhere. One prominent research project that addresses just this is Project Drawdown, which has identified one hundred measures for realistic and effective emissions reductions, listed by the cost-effectiveness of each measure. The list is staggering. According to the researchers, one of the fastest and most effective measures is a change in the production of refrigerators. This is because currently, refrigerators still use gases with a very large cumulative greenhouse effect, which could be completely avoided with comparatively little effort. Alternatives have long been available. Education for women and girls is also listed among the top ten measures. According to the researchers, investments in the global empowerment of women and girls go hand in hand with added value for climate protection grossly underestimated up to now. Among other things, educated women have significantly fewer children, statistically speaking, than uneducated women.[11]

The list of game changers is long. And it gives reason for hope. Educate yourselves and others about the possible solutions, the reasons for hope, the signposts that can already be seen today. We need to start opening our eyes.

4. *Spread the Word*

Few things will challenge social cohesion as much as the Paris Climate Agreement once it is implemented. For that to succeed, we need a public that is better informed than ever before, and that, in turn, requires a new understanding of who has the responsibility to be part of this knowledge sharing: We all do. Of course, the focus is on traditional media and opinion leaders, but it would be foolish to rely on them alone.

Hardly anything touches people more than appeals from people they love. People who read about the climate crisis in the newspaper quickly become resigned to the situation. If you hear about the crisis from a good friend and what they are doing to get involved, you are much more likely to join.

Use your networks, spread the word about what's bothering you about the climate crisis and why it's time for everyone to get involved. Support those people and institutions that contribute to truth telling, i.e., the journalistic, civic, and scientific institutions that are committed.

5. *Educate Yourselves about Each Other*

We call the climate crisis one of the greatest threats to humanity, but we often forget the word "human" inside humanity. This is dangerous, because it leads to debates about the future being dehumanized. Those who oppose the call for radical climate protection often cite "industry," "jobs," or "business locations." Then they say that "technology" will solve the climate crisis and that "the markets" must not be disturbed. But where are the people who created all this? It often seems that in the climate protection debate, the protection of the environment is pitted against the protection of systems. But where are the people who depend on the former and invented the latter just to make themselves feel better? They are missing. Where a debate moves away from people, where it no longer listens to what they really need for a good life, what threatens them and what makes them happy—that's where the debate runs the risk of losing sight of what is essential: Namely, the 7.7 billion individuals living on this planet who all have an equal right to a happy and satisfied life.

11 START DREAMING!

Imagine: Six thirty, time to get up. The sun is out. Small metal containers are lying around in the bathroom next to the sink, containing shampoo, deodorant, toothpaste. Since the packaging reform, which demands that companies guarantee a closed circle of resource use, it has become much cheaper to produce goods with little to no packaging. And since the companies also have to pay for microplastics pollution in the water, hardly anyone is producing plastic packaging any longer. Both regulations fired up creativity in the packaging industry and halved per capita waste within a few years.

I hardly ever have to turn on the heat in my apartment. For a number of years now, houses have been required to install not only fire alarm systems, but also state-of-the-art insulation. Both serve to fight fires—locally and globally.

There is muesli with fruits for breakfast. The products are coming from local organic farms. In community-supported agriculture (CSA), the farmers are guaranteed that their products will find buyers, so they don't have to bow to any price pressures. Since there is no longer any need to pay for marketing and promotion, the products are also cheaper. It is good to know that my food is local, seasonal, and free from chemicals. Most people in my area are already part of a CSA, because it is so easy. The municipal administration has begun to pay for a free month for any newcomers. They can cancel if they like, but only very few do. Most people quickly get used to the comfort of this local, healthy supply of food, and they enjoy not having to stand around undecided in the supermarket.

While I am eating my muesli, I look out the kitchen window. Streams of bicycles are passing by. They are riding on bicycle highways, while intelligent traffic lights, broad streets, and clear signage ensure that the traffic keeps flowing smoothly. Here and there you can still see a car or a bus lane, but they are clearly separated. I, too, get on my bike and ride to work. On the way it begins to rain, and since I don't want to get wet, I stop at the next station, put my bike into the bike parking garage, and jump on the commuter train. I don't need to purchase a ticket, because my pass just costs one Euro per day, and you get it automatically when you register your new address online. Of course, everyone is free to reject it. But here, too, it turned out that hardly anyone does. The financing of the public transit system has functioned smoothly ever since.

All these developments only became possible when the pressure to act began to rise with the water: A few years earlier a great storm had led to the flooding of large parts of the North Sea coast, downtown Hamburg, and several other cities. That's when the Hamburg city council agreed to the ambitious goal of becoming the second European metropolitan area to become climate neutral. After Copenhagen, which had already achieved climate neutrality in 2023, two years earlier. Through the affordable ticket for local public transit, the expansion of bike highways, the ban of combustion-engine cars from the downtown area, and the redesign of parking lots into parks and playgrounds, Hamburg turned from a car-friendly to a bike-friendly city.

Despite initial protests, intense debates, and a failed referendum to stop the initiative, the per capita emissions of CO_2 dropped dramatically within a few months. The great success of the reduction in carbon footprints of households inspired more and more cities to follow suit. Nobody wanted to be the last in the competition toward climate neutrality, which began in the early 2020s.

The communities that received attention nationwide for their creativity and the success of climate neutrality experienced a renewed sense of regional pride. Above all, young people have begun to identify more with their home towns and cities, because they enjoy the prestige of "future fit" local politics. There is a hype for the purchase of regionally produced goods, and exchanges, libraries of things, and repair cafés have become new hotspots.

In the years after the residency reform the price of a square meter stabilized in Hamburg and other big cities. After it was made impossible to own housing without living in it, with only a few exceptions, real estate speculation evaporated. The boom of cooperative housing continues to this day, and in the districts where government subsidies for inclusive and intergenerational living concepts are in high demand, xenophobia has dropped by 50 percent over the past decade.

Things have changed in the working world as well. Ever since the positive "common good" balance sheet that all corporations have to provide with their tax returns was combined with tax reductions, corporate culture has changed dramatically nationwide. After shocking the stock markets initially, the new standard launched a restructuring of emissions-intensive corporations. The pent-up demand in many sectors led to a virtual boom of sustainability consultancies.

Today a workday almost never lasts longer than six hours. All employees, who are often shared owners of the firms, set goals together at the beginning of the year. Their top priorities often include having enough time for oneself, one's family, and friends, to pursue one's interests, to develop new ideas, and to reach as many people as possible with the company's innovations. At the end of the year, profits are divided up into investments into the company, rainy day reserves, developments of new product ideas, and the reduction of everyone's workload.

Most experience it as enriching that they can now dedicate their time to their social life, their children, parents, and friends. When people arrive at work they are more relaxed, and happier. And because companies profit from it as well, more and more firms have copied this model.

There are unexpected side effects: People who now have more time for their families, friends, and neighbors are less often sick, feel better both mentally and physically, and invest more in healthy food and recreation. The profits for the healthcare system are enormous, which in turn leads to higher investments in education and prevention.

A similar effect was seen when the thirty-hour work week was implemented and the basic income of 1,500 Euros was established for all: Unemployment sank to a historic low. Income levels rose since people were no longer dependent on taking poorly paid jobs in order to live a modest but dignified life. Now everyone could take the time to find a profession that really suited them. Overall, the reduction in financial stress led to higher satisfaction.

After climate protection was incorporated as a state goal into the German Constitution, and the idea of coal-fired power until 2038 declared unconstitutional by the constitutional court in Karlsruhe, the new government embraced the Green New Deal. The Deal, as people called it, made headlines as the greatest investment project since the end of World War II. The climate fund, which the government initiated for that purpose, included renewable energy subsidies, a one-time wealth tax, and a carbon tax. It made possible massive investments in infrastructure, support programs for the transformation away from industrial agriculture, and model projects for the carbon-neutral economy. Through investments in parks, museums, and cultural institutions, downtown areas were decelerated and rural areas made more attractive. Of course, the path to all this was not without friction. Large industrial corporations had threatened to move their production abroad if the climate-neutrality law were to go into effect. But apart from a small number who moved their production to Eastern Europe and China, the announcement turned out to be an empty threat. On the contrary: The clear legal framework launched a race for resource-saving innovations, which in turn led to a historic level of investment into climate-friendly production.

Through the introduction of a Europe-wide tax on jet fuel, the railroad system's 25-Euro ticket program, and the EU-initiative for the expansion of a European night train system, the passenger numbers of intra-European flights collapsed. New concepts for compartments, conference rooms, and stable internet connection on trains led to a whole new work culture on the tracks.

In response to protests against the expansion of wind energy and high-speed rail, more people were brought into the planning process. Instead of waiting for citizens to sue, regular Future Forums were established, which people were allowed to attend during work hours. Villages in immediate proximity to wind turbines got to share in their profit, which meant the population now had a vested interest in supporting the sustainable ecological transformation.

The investments of the Deal also improved the situation in the health and education sectors, this being another ecological factor: Most social occupations provide a high societal value and produce very few emissions. It was therefore only logical to raise the value of these occupations in particular, through higher salaries, better benefits, and more job security.

In the past few years after the Council for the Rights of Future Generations received a veto right over all new legislation, the process of legislation has become more transparent and participatory. Government parties use the consultation platforms online, seek contact with speakers of neighborhood councils, and lead studies and surveys in order to meet the Council's evaluation criteria. Since all laws must go through this needle's eye, every ministry has established a department that deals with the long-term ecological and social consequences of new legislative initiatives and legal adaptations.

As emissions dropped rapidly, prosperity increased, which is no longer only measured in income, annual profits, and export balances, but in general well-being. The happiness barometer, as it is called informally, whose publication everyone eagerly anticipates every year, has reached a new record high this year. The increase was small in comparison to that legendary year when they had reintroduced the wealth tax and increased the minimum wage, but the high values justify the government's new course.

Some remnants of the old world continued to exist as well: the big industries, the DAX corporations, and the glass towers where people in suits found it perfectly okay to spend 70 percent of their waking hours running around. And those in big homes with large cars parked in front, who spent large amounts of money on self-discovery trips around the world. They, too, were part of this free society. They were just no longer in the majority.

This is no utopia, nor science fiction. It's a scenario we created together after sitting down for a day and thinking about what is important to us. We afforded ourselves the luxury of calibrating a living environment to be "eco" and "social." By the way, our design is not entirely fictitious. Much of it already exists in many parts of the world.

Although it is impossible to plan the future with certainty, visions are a decisive future driver. Think of Karl Marx: The classless society he prophesied has not yet emerged from the self-contradictions of capitalism and probably never will—yet this vision has been driving people to work for changes in the status quo for over 150 years. On the other hand, the neoliberal utopia of an invisible hand of the market that provides prosperity and justice has also been driving people for decades to adapt reality to the theory through political programs and reforms.

A vision of a different tomorrow can give us the strength to embrace the transformation today. Without this vision, we don't know where to direct our energies. In chapter 3, we deplored the lack of imagination and positive vision that has become widespread in our society. That's why we propose that we must train our imagination, to learn to dream again, to think big and small, outside the box and beyond the horizon. In this chapter, we gather what we think is needed to achieve this.

1. MORAL STRETCHING EXERCISES

Stretch your imagination! And your emotions!

Following the dropping of the atomic bomb on Hiroshima, the philosopher Günther Anders identified a fundamental discrepancy between imagination and invention. Human achievements (inventions) had increased and reached a point of such complexity that one was no longer able to imagine the consequences of the technologies and to comprehend them emotionally.

"Unless all is to be lost," Anders concludes, "the decisive moral task today is the formation of moral imagination, i.e., the attempt to overcome the 'gap,' to measure the capacity and elasticity of our imagining and feeling against the dimensions of our own products and the foreseeable extent of the damage we can do."[1] According to Anders, the means to overcome this gap are "moral stretching exercises." Similar to the stretching exercises in sports, which are supposed to increase physical flexibility and resilience, it is necessary to train one's own imagination and emotions by "stretching." How is this to be done? For one thing, by starting to think in large dimensions. This must be practiced. The climate crisis created by the global aggregate of greenhouse gas emissions is admittedly too large for us to imagine in its entirety. But through practice we can improve our cognitive ability to visualize the consequences of our actions for the world's climate.

On the other hand, we can let ourselves be touched by the disasters that the climate crisis is already causing today. In the parable of the lamented future that we told in chapter 5, Noah manages to convince his compatriots of the impending flood by weeping for future generations. Had the people ignored his tears, the ark would never have been built. If we are emotionally stunted because we do not realize what we are doing with our consumption habits, our energy production, or our production methods, the moral appeals of our contemporary "prophets" will die away without any effect.

Both capacities, imagination and emotional sensitivity, are important not only for understanding our present, but also for the images we form of the future, the positive and the negative.

2. LOOKING BACK FROM THE DYSTOPIAN FUTURE

Imagine the catastrophe! Relentlessly!

Whoever confronts how much we have to lose opens up new avenues for change. This requires the willingness to visualize the apocalypse in all its drama. Only then can we understand both cognitively and emotionally what losses are involved, and why there is an urgent need for action.

David Wallace-Wells's book *The Uninhabitable Earth*, for example, shows what this can look like. It begins with the startling simplicity of the sentence, "It's worse, much worse, than you think."[2]

Stories and images can also help exercise the imagination. Stories like that of poet Kathy Jetñil-Kijiner, whose home in the Marshall Islands is sinking due to rising sea levels. Together with writer Aka Niviâna from Greenland, she describes in the poem "Rise"[3] how global warming is destroying their habitat and traditional ways of life.

When we began writing this book, we asked many people the same question over and over again: What comes to mind when you think of the future?

Many told us about the desire for a stable income, an affordable apartment, a livable environment where you wouldn't have to worry when the kids go out to play. We were confused. After months of reading, discussions, strikes, demonstrations, and occupations of roads, railroad tracks, and coal mines, it was clear to us that weather extremes, economic losses, and climate migration would have a direct impact on our quality of life. But while we were dealing with the disaster that global society was heading toward, many around us worried about their incomes, where they lived, and their neighborhoods.

The amazing thing was that they were thinking a lot about the future and acting accordingly in the present—but the climate crisis played no role. The more we thought about it, the more it dawned on us how big a social task we are facing: We need to understand how closely the threats posed by the climate crisis are intertwined with the individual quest for prosperity and a fulfilled life.

RISE *by Kathy Jetñil-Kijiner and Aka Niviâna*

Sister of ice and snow
I'm coming to you
from the land of my ancestors,
from atolls, sunken volcanoes—
  undersea descent
of sleeping giants

Sister of ocean and sand,
I welcome you
to the land of my ancestors
—to the land where they sacrificed
  their lives
to make mine possible
—to the land
of survivors.

I'm coming to you
from the land my ancestors chose.
Aelon Kein Ad,
Marshall Islands,
a country more sea than land.
I welcome you to Kalaallit Nunaat,
Greenland,
the biggest island on earth.

Sister of ice and snow,
I bring with me these shells
that I picked from the shores
of Bikini atoll and Runit Dome

Sister of ocean and sand,
I hold these stones picked from
  the shores of Nuuk,
the foundation of the land I call
  my home.

With these shells I bring a story
  of long ago
two sisters frozen in time on the island
  of Ujae,
one magically turned into stone
the other who chose that life
to be rooted by her sister's side.
To this day, the two sisters
can be seen by the edge of the reef,
a lesson in permanence.

With these rocks I bring
a story told countless times
a story about Sassuma Arnaa,
  Mother of the Sea,
who lives in a cave at the bottom
  of the ocean.

This is a story about
the guardian of the Sea.
She sees the greed in our hearts,
the disrespect in our eyes.
Every whale, every stream,
every iceberg
are her children.

When we disrespect them
she gives us what we deserve,
a lesson in respect.

Do we deserve the melting ice
the hungry polar bears coming to our
  islands
or the colossal icebergs hitting these
  waters with rage

Do we deserve
their mother,
coming for our homes
for our lives?

From one island to another
I ask for solutions.
From one island to another
I ask for your problems

Let me show you the tide
that comes for us faster
than we'd like to admit.
Let me show you
airports underwater
bulldozed reefs, blasted sands
and plans to build new atolls
forcing land
from an ancient, rising sea,
forcing us to imagine
turning ourselves to stone.

Sister of ocean and sand,
Can you see our glaciers groaning
with the weight of the world's heat?
I wait for you, here,
on the land of my ancestors,
  heart heavy with a thirst
for solutions
as I watch this land
change
while the World remains silent.

Sister of ice and snow,
I come to you now in grief
mourning landscapes
that are always forced to change
first through wars inflicted on us
then through nuclear waste
dumped
in our waters
on our ice
and now this.

Sister of ocean and sand,
I offer you these rocks, the foundation
  of my home.
On our journey
may the same unshakable foundation
connect us,
make us stronger,
than the colonizing monsters
that to this day devour our lives
for their pleasure.
The very same beasts
that now decide,
who should live
who should die.

Sister of ice and snow,
I offer you this shell
and the story of the two sisters
as testament
as declaration
that despite everything
we will not leave.
Instead
we will choose stone.
We will choose
to be rooted in this reef
forever.

From these islands
we ask for solutions.
From these islands

we ask
we demand that the world see beyond
SUVs, ACs, their prepackaged
convenience
their oil-slicked dreams, beyond
the belief
that tomorrow will never happen,
that this
is merely an inconvenient truth.
Let me bring my home to yours.
Let's watch as Miami, New York,
Shanghai, Amsterdam, London,
Rio de Janeiro, and Osaka
try to breathe underwater.
You think you have decades
before your homes fall beneath tides?
We have years.
We have months
before you sacrifice us again
before you watch from your TV
and computer screens waiting
to see if we will still be breathing
while you do nothing.

My sister,
From one island to another
I give to you these rocks
as a reminder
that our lives matter more than
their power
that life in all forms demands
the same respect we all give to money
that these issues affect each and
every one of us
None of us is immune
And that each and every one of us
has to decide
if we
will
rise

Like the farmer from northern Hesse, who did not see himself as a victim of the climate crisis, most people live in the belief that the climate is about "the environment" or the poor people at the poles, in the Sahel, and on the island states, but not about us. Again, the collective imagination fails us to see how our actions today shape the world of tomorrow. It became clear to us that if we don't have images of how our planetary home will change as a result of the greenhouse effect, the fight against emissions will remain a matter for idealists, full-time activists, and professional politicians.

3. IMAGINE!

Learn to read the future! And to write it!

Just as important as looking back from the world of an unleashed climate catastrophe is imagining in detail the vision of a climate-neutral future. If the radical transformation of our material, institutional, and mental infrastructures,[4] as Harald Welzer calls them, is to succeed, we need inner images of what this different, better future should look like. We must learn to imagine the desirable future and to recognize that we can bring it about with our actions.

"The future," says Riel Miller, a futurologist at UNESCO, always exists only in our imagination anyway. But it shapes our decisions and our actions in the present.[5] That's why, Miller says, it is crucial to decide which ideas of the future we want to cultivate.

According to Miller, we learn two concepts of the future from childhood: contingent futures, which arise from external influences, and optimization futures, in which something happens that was planned. What we do not learn, however, are novel futures.

We therefore need a "futures literacy." We need to develop the ability to design and anticipate new futures in order to be able to look at the present from a new perspective. This ability, according to Miller, helps us to perceive the new as less frightening. It sensitizes us to the new that already surrounds us without our noticing it, or that is just emerging. It frees us from the notion of understanding the future as a mere extension of the past. However, the ability to discover and invent must be practiced continuously so that we learn to stay in the uncertainty and to discard our fear of the new. In this way, research, political decisions, and social discussions could be enriched by a view of the future that is independent of calculated probabilities or current habits.

Since 2012, UNESCO has therefore been actively building a worldwide network, the Global Futures Literacy Network, which promotes literacy of the future.

ALEX At some point I had had enough of thinking about societal problems in libraries and stuffy meeting rooms. Together with a friend, I set out to get to know examples of lived alternatives. Concrete utopias, as we called them.

We left the city to meet people who were reviving the countryside abandoned by capitalist progress in order to realize their utopias beyond metropolitan areas. People who had completely withdrawn their labor from "the market" and were instead trying to live in an ecologically sustainable way. We were curious and skeptical at the same time: Are these places of resistance against capitalist society? Are they perhaps even the nuclei of a new society? Or do they only serve as a refuge for disillusioned children of affluence in search of nature and authenticity?

We hitchhiked through Germany, France, and Catalonia to experience the people and places that want to anticipate tomorrow in today. In our luggage were notepads, a recording device, camera, and laptop, because we wanted to tell about these places, about the people and their stories. We wanted to broadcast the answers to our questions to the world—if we were to find such answers. Because this was about our future.

During our six-week trip we visited six different places. Places like the Sozialistische Selbsthilfe (Socialist Self-Help) in Cologne-Mülheim, whose residents describe themselves as "unemployed, homeless, disabled, mentally ill, former drug addicts, and lateral thinkers."[6] Since the 1980s, they have lived together independent of wage labor and social benefits. For them, work is whatever is important to the group; they reflect on their experiences at the Institut für Neue Arbeit (Institute for New Work), which is inspired by the philosopher Frithjof Bergmann.

We visited Can Decreix on the border between France and Spain, a hybrid of self-sufficiency house, urban garden, and open-air laboratory for living beyond the growth economy. It is an attempt to try out the ideal of open, networked, and regionally anchored economies in practice, through closed material cycles and climate-neutral consumption habits; here you can also find the Museum of Useless and Questionable Things, which features, among other things, a plastic bag, a TV antenna, a car license plate, and a refrigerator.[7]

We stopped at the Cooperativa Integral Catalana, a movement organized as a cooperative with the goal of organizing all areas of economic activity and life in Catalonia locally and cooperatively. In the words of its founder Enric Duran, it wants to be nothing less than "a free society—outside the law, state control, and the rules of the capitalist market."[8] To this end, they organize regional economic cycles with crypto and regional currencies, and self-managed health care; they also published a handbook for "economic disobedience." Not far from Barcelona, in the "eco-industrial post-capitalist colony" of Calafou, we met people who experiment with open hardware in the "Transhackfeministlab," who finance their political work by brewing beer, or who, as hackers, try to use open source to provide software solutions against surveillance on the internet.[9]

At our next stop, in Longo Maï, we found an agricultural and political network of cooperatives with over two hundred people living on the principle of a gift-giving and sharing economy. When they moved from Vienna and Basel to the south of France in the late 1960s, they rebuilt a dilapidated farm, abolished both wages and private land ownership, and founded a network that today includes farms and initiatives in France, Germany, Austria, Switzerland, Ukraine, and Costa Rica.[10]

At the Grandhotel Cosmopolis, the last stop on our trip, we dropped exhausted onto the beds of the Utopia Room. The house is a "social sculpture as an intercultural meeting place in the old town of Augsburg," a mixture of hotel, asylum shelter, and cultural center. Some hotel guests who have found asylum in Germany live there permanently and participate in the hotel and cultural activities; we, as guests without asylum, stayed only for a few nights and were amazed at the many offerings in the *Bürgergaststätte* (restaurant by all for all), the studios, and the café-bar.[11]

Our trip was an eye opener. A trip through the diversity of lived utopias, with all their hopes and solutions, but also shortcomings and contradictions. They showed the potential of visionary thinking and also at which points the implementation can fail due to everyday and ideological differences. To this day, the impressions from that trip are a treasure chest full of connections to what is already available to us today on the way to a just, post-fossil age. What we need is a political majority willing to unearth this treasure.

4. THINK UTOPIAN

Draw Utopia onto your cognitive maps!

Riel Miller's ideas point to the potential of something we want to call "utopian thinking." That will be crucial if we want to shake off thinking in terms of apparent constraints, quarterly targets, and legislative sessions. Only then can we open our eyes to paths that are different from the ones we already know. Utopian thinking frees us from the disastrous habit of doing things "because they have (apparently) always been done that way" or "because there is no other way" (thus blocking the way for ourselves to do things differently). If the future is to be different, we need utopias.

In 1891, Oscar Wilde wrote: "A map of the world that does not include Utopia is not worth even glancing at, for it leaves out the one country at which Humanity is always landing. And when Humanity lands there, it looks out, and, seeing a better country, sets sail. Progress is the realization of utopias."[12] We need imaginary places of a better tomorrow in order to successfully overcome existing grievances. It is not about the blind hope for a rosy future that would make the suffering of the present more bearable and cloud our ability to recognize current problems. Rather, utopias are the magnet that attracts people to the future so that we do not wither in stagnation. If we turn the perspective around and look from a desirable world of tomorrow back to the present, we can better identify paths that lead to that tomorrow. Anyone interested in such a change of perspective will quickly come across Futurzwei.Stiftung Zukunftsfähigkeit, which practices just that. It regularly publishes so-called "stories of success" in a "Future Almanac" to visualize examples "of good ways of dealing with the world."[13]

If you look further, you will find examples such as the transition town movement,[14] the online platform Future Perfect,[15] or the places mapped by environmental activist John Jordan in his book *Paths through Utopia*.[16] Other examples are the "real utopias" collected by sociologist Erik Olin Wright[17] and listed by Right Livelihood laureate Anwar Fazal in his *Sourcebook for Changemakers*.[18]

These are places and initiatives where we can find the future already today, examples that can inspire us because they show how living together and doing business based on a different relationship with nature and more active political participation can work. We can already see the cracks in the concrete of current power structures: Countless initiatives have created new social realities that anticipate the world of tomorrow in practice. Their methods owe much to the imagination and the courage to try out something new. This is how a new concept of a society that is fit for the future becomes imaginable.

If we want to bring about the end of the climate crisis, we must learn to close the gap between production and imagining. We have to learn to imagine the connection between our actions and their consequences, and to comprehend it also on an emotional level. The risks for other regions of the world, future generations, and nature must be considered in the construction of energy supplies, in economic models, and in technological development. A look back from the postapocalyptic future, future literacy, and the development of positive visions are the first points of reference for this. If we train our moral imagination, we may be able to overcome the lack of collective leadership that was a cause of past catastrophes such as the financial crisis and the Fukushima disaster—and which today threatens to destroy the very basis of human life on earth.

12 GET ORGANIZED!

Why did we call this book *Beginning to End the Climate Crisis*? We named it that because we know that, from a purely scientific point of view, it is possible to get this crisis under control. This possibility is called the 1.5-degree target. The probability of still achieving this target is minimal. But it is there. And as long as science confirms this, it would be grossly negligent not to do everything in our power to achieve it.

The good news is that science not only knows that the Paris Agreement is feasible. It also has a very concrete idea of how this can be done. The IPCC, the Intergovernmental Panel on Climate Change, has developed various scenarios for the next few decades that show how global warming can be limited to prevent the worst damage.

The bad news: Right now, there is no reason to trust that this will happen. Why should action be taken all of a sudden, after nothing was done over the last thirty years? In the past, too, there were international climate negotiations, declarations of intent, and national targets.

Political actors and decision makers are the ones who should initiate this change. They must set emissions targets for their respective countries and create a regulatory framework that guarantees compliance.

But this does not seem to work: So far, the costs of such measures have been greater than the incentives. On the one hand, governments were faced with powerful economic players who had little interest in strict regulations, carbon taxes, or other restrictions. From their point of view, this is logical. These companies have to defend themselves against competition from home and abroad, increase their sales, and preserve jobs. Saving the climate is not one of their business objectives. On the other hand, governments fear the people because ambitious climate protection hasn't helped anyone win an election. The Green Party's modest election results up to now are proof of this.[1] There are no sanctions for noncompliance in the Paris Agreement.

LUISA For me Annegret Kramp-Karrenbauer[2] was the straw that broke the camel's back. That was in December 2018, but it was not the reason why I became part of Fridays For Future. To explain that, I have to elaborate a bit. Let me start at the beginning.

Ever since I had learned about the phenomenon of a warming planet, I had been an advocate for climate and the environment. I became interested in the issue and became active, both inside and outside of school. As time went on, I became more and more bewildered by how apathetic political decision makers were to the drastic warnings of climate scientists. And because the mountain of questions kept growing, a year after graduating from high school I decided to study geography. In December of the same year, the Paris Climate Agreement was adopted. I was nineteen years old and relieved. The governments had gotten their act together after all. That would probably be reflected in the emission graphs. From time to time, I gave speeches, wrote articles, got involved in various projects, and was a devout vegetarian. Just like many others.

Only in retrospect do I understand that my commitment was connected to a certain level of comfort: I thought what I was doing was important and I enjoyed it. But what if I stopped doing it? The world would not come to an end. I was not a full-time activist and had no qualms about flying to London several times a year. Despite all the criticism I had of our government, a certain trust remained. Somehow, I hoped, it would be able to cope with this crisis, as it had with so many others before. It would not have to be with flying colors. You don't sign an agreement like the one in Paris lightly, I thought. Surely things would get better now. And I guess I was not the only one who had these expectations.

A series of events that occurred in 2018 fundamentally changed my view of things.

Two years after the Bundestag unanimously approved the Paris Agreement, German Environment Minister Svenja Schulze declared that Germany would not only miss the climate targets it had set itself for 2020, but would miss them by a wide margin. Back then, a heat wave had just hit Germany; drought reports dominated the news, and forests were burning in Brandenburg. And as it turned out, the statement by the environment minister was not merely an announcement of a technical correction or an admission that the targets had been miscalculated. A study commissioned by Greenpeace and conducted by the Fraunhofer Institute showed just a few months later that it was perfectly possible to meet the climate targets that the government had set itself. It would just be time-consuming and expensive. The environment minister had thus openly acknowledged her lack of political will. She had made a declaration of political bankruptcy. I felt infinitely naïve, and more than that, I felt cheated.

The climate crisis is not like an untidy room that takes more time to tidy up if you haven't done it for a while. The climate crisis is like a house on fire. Every minute you wait to put it out, it becomes less likely that the worst damage can be prevented.

I've never been one of those people who accuses government members in general of being bad people. Or of being for sale. But the idea that at one point a deliberate decision was made to abandon climate goals simply because it was opportune or convenient—that idea left me feeling that my own government had decided against me and my future. At that moment, my deep conviction that the climate crisis would be dealt with was shaken.

Meanwhile, Germany continued to be hit hard by the climate crisis. By the end of the summer, there had been a total of 1,708 fires, four times as many as the previous year. For weeks, the Rhine could not be navigated, fields were bare, and prices for potatoes and onions rose steeply in stores. People became ill; in Berlin alone, nearly five hundred people died from the heat. That summer was as deadly as almost no summer ever before.

For the first time, I was afraid of the climate crisis, of what would await us tomorrow, what would become of my life and the lives of the young people around me if governments like ours continued to allow all this to happen. It was the first time I wondered how anyone could bear the knowledge of this permanent, wanton destruction at all.

When the opportunity arose to travel to Poland for the climate conference, I didn't hesitate. Were there countries, I wondered, that took the climate crisis more seriously than Germany did? In freezing December I traveled to Katowice, a small mining town where the air tastes strange because coal dust drifts through the streets.

There I learned that since the conclusion of the Paris Agreement numerous additional coal-fired power plants had been built. And that the construction of more than one thousand additional coal-fired power plants is still planned worldwide. More than one thousand coal-fired power plants!

Each one of them will have to run for several decades before the investment pays off. If this were to happen, the 1.5-degree target would be dead. People around me were taking notes, scrolling through the calendar of events, already on their way to the next event. My heart sank.

Later, an event with Greta Thunberg was taking place in the same room. The room was only sparsely filled, the moderator a disaster. When it was over, I went to Greta and offered her my support. She seemed to me to be one of the few at this conference who was paying attention to all the madness. On Friday of that same week, I went on strike for the climate for the first time.

And then Annegret Kramp-Karrenbauer came along. She was elected chair of the CDU just a few hours after our first strike. After fourteen years of Merkel, I was stunned that a woman who had never shown the slightest interest in the climate had been elected leader of the country's most powerful party. The leader of the party that would presumably provide the next chancellor. And that was in 2018, so it looked like everything would stay the same. Unless we get louder, much louder.

SORRY, I DON'T HAVE TIME TO PROTEST

When we were telling people what we were writing about, many expressed the wish for as many tips as possible for a climate-friendly life. Maybe one day we'll write a book like that. But first we need everyone to change the framework conditions so that a truly climate-friendly life becomes possible in the first place.

Because it is not possible today. Imagine social conditions as a large empty space. In that space people want air they can breathe, which should be a matter of course. The only catch is that over the past two hundred and fifty years, the space has filled up: with highways and industrial sites, with factory farming and monocultures, with coal-fired power plants and pipelines, airplanes, and lots of old houses heated by equally old oil-fired heating systems. All of it sinks into the soil, pollutes the air, and makes people sick. What's left, in this big space, is the square meter of meadow in the middle, where we can still have our plastic-free, vegan party now. That may feel great at the moment. But it ignores the elephant in the room, it doesn't change the conditions. And if you don't change the conditions, if you don't organize, if you don't see yourself as part of the critical mass that collectively has the power to reorder the space—you can try as hard as you want in private. It will not be enough.

Therefore, this book does not explain how to make shampoo yourself and how to travel in a climate-friendly way (both are worthwhile and can be read about in books). We are concerned with something different: If we are serious about creating a world that does not reach further tipping points and is nourished by what our planet provides in terms of resources and ecosystem services, we must change our exploitative way of living. And in fact, we need to do that on a very large scale. We need to embark on the biggest transformation since the industrial revolution. Some people talk about *Agrarwende* (transformation of the agricultural sector), *Verkehrswende* (transformation of the transportation sector), and *Energiewende* (transformation of the energy sector).[3] We are talking about a climate revolution.

Revolutions do not fall from the sky. They need the pressure of the masses. If there is no pressure, things will continue as they have in the last three decades. Think tanks such as Agora Energiewende,[4] institutions such as the German Advisory Council on Global Change,[5] numerous environmental associations and research institutions have long since pointed out what would have to happen for Germany to at least start doing its part to alleviate the climate crisis. The concepts have been around for a long time. What has been lacking, however, is the political and social will.

Those who embody this will today are the young people who, inspired by Greta Thunberg, are taking to the streets. But it doesn't have to stay that way. Nor should it. The ability to develop an awareness of the climate crisis does not depend on age or generational affiliation. It seems that the line of conflict is much more likely to run between those who espouse the status quo because they believe they benefit from it, and those who dare to question it, because they are willing to set overriding priorities.

ALEX I experienced a personal turning point in September 2018. The brick buildings in Cambridge shone in the late summer sun that day, barges with tourists meandered through the water on the canals. Couples took photos in front of the Bridge of Sighs, others strolled across the green of the English lawn. For ten days, at the University of Cambridge, we spoke with scientists, who were studying global environmental change at the university, the United Nations, or other institutions. Other working groups at our Future Academy discussed populism, innovations, mobility, and utopias—and we talked about the climate. We learned how tree-ring analyses could be used to study the climate thousands of years ago. That the Bramble Cay mosaic-tailed rat was the first mammal to become extinct in 2016 due to climate change. And how global warming is affecting Indigenous communities in the Arctic Circle. We discussed the potential of geoengineering to solve the climate crisis, we visualized ocean acidification and the consequences of ecological tipping points for the world's climate.

I learned a lot, but already on the first day I felt like I was in the wrong place. While we were discussing the scientific background to the climate crisis, the police in Düren began to clear the Hambach Forest. The tree houses were to be demolished, the trees cut down, and the activists driven out of the forest. For several years, they had occupied the trees next to "Europe's biggest hole," as someone had called the open-pit lignite mine operated by RWE in Hambach. They wanted to protect the forest from the greedy shovels of the coal excavators.

Now the mine was to be expanded so that the lignite under the forest could also be burned. I was stunned. This was only the first day of the Future Academy, but this much I already understood: If we wanted to achieve the goals of the Paris Agreement, 80 percent of the world's remaining coal had to stay in the ground.

In a mixture of anger and helplessness, I asked for the microphone at the evening event. "As you have surely noticed, the eviction of the Hambach Forest has started today," I began in front of the academy participants. I was speaking English, and my voice did not sound as firm and determined as I would have liked. It was the first evening, I didn't know anyone, and I wasn't practiced in mobilizing people for political causes. Only a handful of the almost one hundred people showed up at the meeting place I had suggested for planning a solidarity action.

Did they not understand how big the threat of greenhouse gases really is? Did they not know that RWE power plants were among the largest sources of CO_2 in Europe? Did they trust that the German coal commission would reach a sufficient agreement? Did they believe that the protest was pointless, or did they just not care?

There I was sitting in a future academy in one of the most renowned universities in the world, together with a hundred people described as "elite" by others or even themselves, wondering why only five of them were paying attention to one of our central future tasks.

That first day in Cambridge was the day I lost the sense that I needed to know more in order to be able to act. In a place where the world's knowledge gathered, a simple incident was enough to lure me out of my comfort zone. Because I knew the numbers and stories about the climate crisis and took them seriously, a headline was able to stir me up. Because I let myself be touched by the threat of coal excavators to a forest, I realized how real the dangers of the climate crisis are. I understood that the exploitation of nature by our fossil greed is very concrete in the here and now, and not far from where we live.

Three weeks later, I stood with fifty thousand people on the edge of the Hambach Forest. It was the largest demonstration that had ever taken place at the Rhenish open-pit mine.

WHY ORGANIZE?

Imagine two scenarios.

Scenario one: A single person gives up meat in order to protect the climate. A good step, but on the whole, it achieves little. This one additional vegetarian is initially silent. But she is not alone. She convinces someone else to join, and that person does likewise, so the number slowly grows. There are about six million vegetarians in Germany today. Supermarkets are offering more and more vegetarian dishes, and restaurants are also offering more and more vegetarian dishes. Are emissions in industrial production decreasing? Hardly. Are ministries taking this as an opportunity to quickly draft ambitious legislation to reduce emissions? No.

Scenario two: Thirty million people across the country boycott meat for a month to protest the conditions of its mass production. This would be an unprecedented, coordinated consumer strike. There would be outcry, debate, media coverage, and political pressure. While people would basically be doing nothing more than going vegetarian for a month, this strike would be highly political and would have a very different effect on meat producers, who would now fear being permanently stuck with their products. They would have an interest in ending the boycott. That would be historic.

That is the magic of organizing and scaling, applicable to any context.

One person who decides not to fly remains a single, often even somewhat helpless believer. But if a hundred people organize to peacefully occupy major aviation industry hubs on a weekly basis, they can spark a debate. A child who leaves the classroom in geography class because he or she wants to protest the fact that not enough is being said about the climate crisis remains a child who has skipped class. Thirty, three hundred, or three thousand kids who organize and walk out of their geography classes have the power to change the curriculum. One small town deciding to become carbon neutral by 2025 remains invisible. One thousand small towns joining together with the same goal—that is the beginning of an urban revolution.

3.5 PERCENT

It's a matter of organization and mobilization: Any action, no matter how small, can have a big impact if it's launched at the right moment, with the right narrative, and by as many people as possible. In the past—think of Mandela, King, and Gandhi—people brought about previously unimaginable changes in this way. And they did so peacefully. Movement researcher Erica Chenoweth has studied what has made social movements successful.[6] Movements that were strictly nonviolent were twice as successful as those that were violent. Not a single nonviolent movement failed once more than 3.5 percent of the population was mobilized. That is not a small number; in Germany, it would mean 2.87 million people. But it is also not utopian.

What we urgently need is a major, socio-ecological transformation. The period of time we have to achieve this is very short. The moment for big mobilization is now. That's why we're calling here for scalable, nonviolent action, and why we're telling everyone to organize.

If the will is there, the question is how. We sketched out six essential aspects below.

1. Discover the Why

On August 28, 1963, 250,000 people flocked to Washington to hear Martin Luther King Jr. speak. They had not received an invitation, there was no web page, and Facebook Events had not yet been invented either. But they came and heard King talk about his dream. Why did all these people come? They came because King had found a clear language for the suffering and injustice of segregation that they had to live with every day. Because he articulated a vision that gave them hope for a better tomorrow and showed effective ways to make it a reality. King and his actions were the embodiment of a future they longed for. But in the first instance, it was not about what he did, but why he did it. It was this "why" that inspired many people: his belief that all people were created equal by God and that they should live in peace with each other. That was the conviction they embraced.

If King had chosen that day not to protest but to sing, they probably would have sung along; if he had announced resistance through silence, they would have been silent with him. When the "why" is answered, the "how" and the "what" are secondary.

In the climate issue, the powerful, engaging answer to this why has been missing for a very long time. Why should one commit oneself to fighting the climate crisis, if this crisis is not even visible? When you yourself are not even affected? When its devastation will be felt at the other end of the world or only in the distant future? Why should we get involved in solving a problem that we did not cause ourselves? It is easy to say, "Let's stop climate change." It's much less easy to explain why we are all called to do so.

But then came the summer of 2018, and with it came Greta, who answered the big question of why with stunning clarity: Science says the climate crisis is here. You adults who are not complying with the Paris Agreement are robbing us children of our future.

That's why we are on strike until you act. Why should we study for a future that will soon no longer exist? This why is the narrative of the school strike. It mobilized the masses. For the first time, the global injustice of the human-made climate crisis was captured so succinctly that young people around the world took ownership of the fight against this injustice.

It all starts with this why. Movements that want to make a difference need an answer to that question. Why does one want to get involved? What is the inner motivation, the deep conviction? Only on that basis can a narrative develop that attracts people, a cornerstone on which to build everything else.

2. *Open Your Eyes*

Once people realize that it is up to them to demand change, once they know that they can take the story into their own hands, then anything is possible. In Germany especially, people should know that. In 1989, we experienced how a peaceful revolution can work. What power people can unleash when they unite. They can tear down walls and overthrow governments.

"Power," writes philosopher Hannah Arendt on the subject, "corresponds to the human ability not just to act but to act in concert. Power is never the property of an individual; it belongs to a group and remains in existence only so long as the group keeps together."[7] Since the power of many individual actions is not enough, we must join together to make a difference. Only groups are powerful and their power increases as the group grows.

It is tempting to think of the great names in revolutions, to admire them, and to get lost in the radiance of their action. It's easy to let yourself off the hook; not everyone can be King, Gandhi, or Mandela. Fortunately, no one has to be.

We should learn from historical figures and movements what is possible when people really want it. By the way, the few icons of collective memory did not do their work alone either. Their stories need to be complemented with the stories of those who stood beside them, in front of them and behind them, and made the world what it is today.

We need narratives about those who have made the seemingly impossible possible. Narratives about historical figures like Alice Paul, who played a critical role in the fight for women's suffrage. Or about Lech Walesa, who went from being a Polish electrician to a civil rights activist, president, and Nobel Peace Prize winner. But we also need the stories about people like the painter Katrin Hattenhauer, who played a key role in the peaceful revolution in the GDR and fought with many others for a free country, or the entrepreneur Heinrich Strößenreuther, whose commitment in 2016 made him a decisive force behind a new Berlin mobility law.

Open your eyes. Learn about the stories of those who have made history, big and small. Let them encourage and inspire you, strengthen and empower you. They can't tell us or you what exactly needs to be done. They can't tell us how to solve the climate crisis. But they can tell us what attitude it takes to get started wherever we can. And how to approach the big questions, step by step.

In Kenya, the country's first planned coal-fired power plant was prevented from going ahead in June 2019, because the local population had mobilized against it.[8] In a country that has no experience whatsoever with coal-fired power and in which many people are not even connected to the power grid, this is a remarkable emancipatory act. Imagine such an act in the context of global solidarity, imagine communities around the world that do not want to be driven into dependency by fossil megacorporations. If the privileged and the deprivileged would work together against the old ideal of prosperity—imagine what all would become possible in the struggle for a better future.

3. *Team Up and Look Out for Each Other*

Get together, online and offline, locally and globally. A single person demanding uncomfortable change will remain a lonely, uncomfortable person. A hundred people demanding uncomfortable change are hard to ignore. The vested interests are skilled in the strategy of sitting out, watering down, waiting for the media attention to turn to a new issue. What we need is staying power. The activists behind the bicycle referendum in Berlin did not stop mobilizing after they handed over their signatures. They are still organizing today and putting pressure on the implementation of the *Verkehrswende*. The civil rights movement in the United States also did not dissolve when politicians first offered a compromise. The civil rights activists remained persistent and determined in their demands. Only because they were organized could they keep up the pressure long enough.

Those who surround themselves with people who are working for the same cause can also better empathize with those who are suffering. The climate crisis not only makes people sick, it also overwhelms them. It can only be countered with empathy and attentiveness. For example, by supporting those for whom everything has become too much. We can learn from people like environmental activist Joanna Macy, who has shown us for decades how to transform resignation into empowerment, how to reconcile our own needs with political commitment.[9] So get together, get active—but take care of each other as well. For a group to have staying power, it mustn't run out of steam.

4. *Copy from Each Other*

If you want to organize to make a difference, you don't have to reinvent the wheel. It's enough to apply the knowledge that's already out there to the situation at hand. This is not always easy, because this knowledge is rather insufficiently documented. The techniques and dynamics of coordinated assemblies, nonviolent action, and social movements have long been neglected in research, and they appear even less frequently in school curricula. While most people may have a vague idea that Martin Luther King Jr. played a crucial role in the civil rights movement, few know what triggered his involvement, how he and his many supporters planned their strategy, how they mobilized, and what hurdles had to be overcome along the way.

In his book *From Dictatorship to Democracy*,[10] political scientist Gene Sharp, also a recipient of the Right Livelihood Award, lists 198 methods of nonviolent action: 198 ways for individuals, groups, or institutions to act. All of these methods have been used in the past, and are thought to have had a decisive impact on numerous revolutions, such as in the Arab Spring, in Serbia, and in Ukraine. Sharp was meticulous. Over many years he researched the history of nonviolent action and especially the strategies employed by Mahatma Gandhi. Every conceivable form of action can be found in the list. It is sorted according to nonviolent protest and persuasion (such as public speeches, leaflets, singing), social, economic, and political noncooperation (consumer boycotts, rent withholding, election boycotts), and nonviolent interventions (sit-ins, guerrilla theater, overwhelming bureaucratic systems). Number 62 on the list is "strike by students or schoolchildren."

Many think of protests, marches, and vigils when they think of organized actions. But the spectrum of possibilities is so much wider. We just need to make use of it.

The critical factors are, 1. scalability, i.e., the possibility of implementing actions in both small and large formats; and, 2. visibility, which is created by consciously using public spaces and media spaces and making them our own. Activism today quickly runs the risk of becoming clicktivism. But the shift to the web gives away a great potential, because a certain relevance is created by physical presence alone: If thirty thousand people sign an online petition, it is quickly forgotten; but it inevitably attracts attention if only three thousand become publicly active. Finally, and this is the critical point, it is about repoliticizing what happens in the private sphere and remains largely ineffective there. About repoliticizing what people do when they "start with themselves."

An example: You consciously ride a bicycle and are annoyed at the fact that there are no bicycle lanes? It's not just you. Years ago, people organized and got together to visibly occupy public spaces for cyclists. They formed a "critical mass." Through various actions, bicycle activists have already achieved a lot. In the Netherlands for example: When urban car traffic there became increasingly dense in the 1970s and more and more children got run over, activists simply painted bicycle lanes onto the streets. Such actions of civil disobedience have led to the fact that today the whole world knows about the high quality of life in bicycle-friendly Amsterdam. But less disobedient actions can also have an effect: After Heinrich Strößenreuther and his supporters brought about the successful Berlin referendum on bicycles, paving the way for a bicycle-centric mobility shift in the city, people elsewhere followed suit, and similar referenda were later held in fifteen other cities.

In Berlin, there were sit-ins and vigils for dead cyclists; district groups networked with each other, and people put pressure on politicians with well-thought-out traffic concepts and petitions. Heinrich Strößenreuther's example shows what Gene Sharp also emphasizes in his book: What you need is a strategy. What established environmental associations had failed to achieve over decades—a change of direction in the capital's mobility policy—the *Volksentscheid Fahrrad* (bicycle referendum) accomplished it in just three years, thanks to an intelligent strategy.

When Tarana Burke used the phrase "MeToo" for the first time in 2006 to talk about sexual harassment, there was no big outcry. When actress Alyssa Milano used the same hashtag again eleven years later it was a completely different story. Within twenty-four hours, #metoo was shared half a million times on Twitter. This triggered an earthquake in Hollywood and led to a global debate about sexual abuse.

In March 2017, when the famine in Somalia had reached disastrous proportions, influencer Jérôme Jarre made public that there was one commercial airline still offering regular flights to Somalia. That was Turkish Airlines. Jarre mobilized his network of key social media influencers, including Casey Neistat and Ben Stiller, who now rallied behind the hashtag #lovearmy. In a short video, Jarre told the story of the famine that no one was talking about. And urged his fellow influencers with the hashtag #turkishairlineshelpssomalia to pressure Turkish Airlines to provide a cargo flight to Somalia. At the same time, #lovearmy launched a crowdfunding campaign to raise money for food and water.

LIST OF NONVIOLENT, DIGITAL FORMS OF ACTION

When Gene Sharp published his list of nonviolent forms of action in 1973, the internet was still in its infancy. Since then the web has increased the repertoire dramatically, of course. In order to give insight into this spectrum, we have collected thirty-four exemplary methods that can be used to organize or mobilize in the digital age. Some methods are not legal, partly for good reason. But as Sharp writes, when in doubt, any nonviolent means can be legitimate for exposing or overcoming injustice. And without digital disobedience, a society runs the risk of disempowering itself on the worldwide web. Had Edward Snowden and Chelsea Manning not broken the law, we would still be in the dark today about the Orwellian levels of surveillance or the war crimes committed by US soldiers in the Iraq War.

1. start online petitions
2. reject or approve via email
3. flood online consultations
4. write mass text messages
5. targeted tagging of people on content
6. strategically follow and unfollow on social media
7. memes
8. graphics
9. trending hashtags
10. protest through profile pictures
11. appeals in bios
12. social media challenges
13. share written or spoken speeches
14. chain messaging on WhatsApp, etc.
15. use Facebook Events to call for action
16. mass requests for online forms
17. cancel online
18. boycott of apps
19. boycott of accounts
20. boycott of websites
21. boycott of online service providers
22. reporting accounts
23. fake accounts
24. satirical accounts
25. hacking
26. viruses
27. whistleblowing and leaking
28. use of prohibited websites
29. sharing prohibited content
30. redirecting IP addresses
31. open-source information dissemination
32. open-source software distribution
33. crowdfunding
34. crowdsourcing

Within hours, the airline responded by providing two cargo flights to Somalia, while the crowdfunding campaign raised several million dollars for food. This allowed NGO staff to distribute the donations locally, while other aid was purchased directly in Somalia to strengthen local markets.

The #lovearmy campaign has not yet grown into a powerful, globally organized movement. However, it shows what is possible when the right methods are used at the right moment. Moreover, this was also about telling a story in such a powerful way that people want to become a part of it.

Luisa: I met Jérôme Jarre when we were waiting together for Barack Obama. With some others, we had been invited to meet him during his European tour. Rarely have I seen so much fuss made over a single person. Dozens of security guards had cleared our bags out of the room, selfies were forbidden. We waited a full hour for the American ex-president. Meanwhile, Jérôme told me his story, beaming all over his face. He had big plans for his #lovearmy and he was sure he could achieve his goals. People who fill the internet with life have an incredible amount of power. At least in theory. You just have to make them aware of it. And you have to show them how to use this power.

When Barack Obama entered the room, everything seemed to vibrate. He still looked eerily presidential. Obama looked around the room and said something like, "Wow, what energy in this room." He certainly hadn't said that for the first time. But I thought of all the opportunities we had and agreed with him.

5. *Come to Stay*

The stories of nonviolent resistance are rarely talked about. What does stick in the collective memory? Maybe Gandhi and King: one who sprinkled salt and rallied hundreds of thousands behind him, and one who told about his now world-famous dream. Hardly anyone knows the stories of how many people before them tried without success what these two succeeded in doing. Nor do they know the names of all the courageous people who stood up to injustice with their bodies, who were insulted or beaten up because they rebelled. Today, their actions are considered heroic; back then, they were arrested for them. We should remember this when people are on trial today for occupying trees or coal mines.

We have a lot to do, and it will not be easy. We will have to make many attempts, even unsuccessful ones. In the end, it will be about being in the right place at the right time, and for that we also need luck. Trial and error. That's another reason why many more people have to join us. With each individual, the chance of a breakthrough increases. We are on strike until you act.

6. Make Demands of Those around You

LUISA I have already told you that I had never organized a public demonstration before our first strike. A short time later, I was faced with the next challenge: In the long run, we needed reinforcements in Berlin. A local group that would organize the strikes together with me and the others. At this point, I had already spent weeks trying to win people over to our cause, and many of them signaled their general willingness. But when it came to taking responsibility, a lot was left to me and a few comrades-in-arms. We offered people less extensive fields of activity. We hoped that more people would support us if the to-do lists were not so long and less time-consuming. To no avail. Then we turned the tables. I had remembered Bernie Sanders's campaign for the 2016 presidential primary. At the time, his campaign team mobilized using such a successful technique that books have since been written about it. People were invited to town hall events. But instead of passing around sign-up sheets for email addresses at the end, as was customary, and then waiting in vain for attendees to respond to the follow-up email, they were asked to sign up that same evening if they wanted to host a fundraising dinner, for example. Those who did so were registered directly for the campaign. That's how they recruited an unprecedented number of volunteers. Not with friendly emails, but by radical delegation of responsibility.

Inspired by this, after moderately successful weeks, one day we simply founded a Berlin organizing team. Just like that. During the strike, we asked people to get in touch if they wanted to support Fridays For Future. And before they could change their minds, they were added to the relevant WhatsApp groups and given a task. My expectations were greatly exceeded.

Since then we've seen it every week: people surpassing themselves in working together.

Nothing is easier than underestimating people. Nothing is more depressing than "making an effort" that is limited to one's private life and then drowns in the maelstrom of climate catastrophe. Nothing motivates more than the knowledge that one's own contribution is important.

That is why we organize. On all levels.

Forget Meat-Free Monday. Start a Meat-Free Month. And not just with your best friend, but with the whole department. Make it known. Then ask the whole company to do it, and then the competition. And finally announce that you are the first group of companies that has joined forces to demand an end to the cheap meat craze. And so on.

The end of the climate crisis becomes possible when people, companies, and institutions join forces. When they coordinate and become more and more; when they admonish the public to finally do what needs to be done. Great ideas and concepts have been around for a long time, and there are more of them every day—they just need to be implemented.

A few million young people not going to school or university caused an unparalleled stir in 2019. Imagine what would become possible if all others who want to be part of the story of the end of the climate crisis start organizing as well?

The answer is: Everything.

EPILOGUE

When we started writing this book, we were sure that it was possible to avert the climate catastrophe. This view was shared by the Intergovernmental Panel on Climate Change—and we, the possibilists, saw this as the starting signal we needed in order to act.

Then a lot of things started to happen before we had even written a single chapter. As if 2019 wanted to prove to us how many catastrophes can occur in just a few months. In Germany, April turned out to be the thirteenth abnormally warm month in a row—something that hadn't happened since 1881. No sooner was it over and the United Nations declared that one million species worldwide were threatened with extinction—a whole new dimension of what was long known about extinctions of species. We had just written two chapters. Then came June. It became the hottest June worldwide since record keeping began. In the Indian city of Chennai, water reservoirs dried up, causing five million people to suffer from water shortages for weeks. The heat continued relentlessly. At the end of July, an incredible twelve billion tons of the Greenland Glacier melted within twenty-four hours. The United Nations declared that environmental degradation was now responsible for a quarter of all illnesses and premature deaths.

Before we had written the last chapter, people were wantonly setting fire to the Amazon rainforest. Never before had there been such a fire. Pictures went around the world of places that once belonged to the lungs of the world. Now they are just burnt stretches of land.

In some moments, one is dominated by the belief that we can no longer be saved anyway, and that the "end of the climate crisis" is nothing more than a fairy tale, as optimistic as it is improbable.

But, to our own surprise, we have experienced just the opposite in recent months. While we received disaster news after disaster news, bit by bit the other side of the coin appeared before our eyes. From one month to the next, the largest climate movement in the history of humankind was formed. It managed to put the climate crisis on the political agenda. We demonstrated and researched. And in the process, we have come to understand more and more precisely what is feasible if only enough people join forces. Because all the people on the streets have understood that otherwise their own future will be taken out of their hands. We have heard from the pioneers of this movement, from people who have long since embarked on the great transformation, some of them very successfully. We have learned that the biggest barriers are not technology, the market, or money, but the hurdles in people's minds.

These hurdles must be understood, analyzed—and overcome. That is the purpose of this book. We don't just know of the existence of the climate crisis, we also learned about its causes. Many people have understood that alternatives are necessary. They are putting them into practice every single day. When you read their stories, all excuses, constraints, or complaints about the lack of alternatives that characterize the status quo seem increasingly absurd. For possibilists like us, those lived alternatives are a source of strength and tools of the trade. The only question is: Will we be fast enough?

August 2019: A glacier is being buried in Iceland for the first time in history. It is a victim of the climate crisis and has been named Ok. There was a ceremony in its honor, and the writer Andri Magnason authored text for a memorial plaque. Maybe our children will read it, maybe our grandchildren or great-grandchildren: "A letter to the future. Ok is the first Icelandic glacier to lose its status as a glacier. In the next 200 years, all our glaciers are expected to follow it. This memorial is meant to commemorate that we know what is happening and what needs to be done. Only you know if we have done it."

Whether the end of the climate crisis can begin now is up to us. Whether the big changes will come by design or by disaster is up to us. We have taken on something unprecedented.

Our advantage is that we know what must be done. We also know how, and above all we know that it is possible.

This is our chance to write the history of our future.

Change is coming.

ACKNOWLEDGMENTS

Writing a history of the future is teamwork. We therefore thank Tom, Julia, and the whole team at Tropen and Klett-Cotta; Katrin, who helped bring this book into the world together with us; Ulrich and Ronald for their support and encouragement in the making of the book; Stefan Rahmstorf for his critical look at the scientific data and sources; Ole, Kajsa, and the Right Livelihood Foundation, for making all this possible in the first place. For the wonderful work with the English translation, a huge thanks to Sabine von Mering.

LUISA My thanks go to my family, siblings, mother, and grandmother, who have supported me in all the craziness, and to Daniel, whose energy and inspiration have led the way. In greatest solidarity with those who I have organized, mobilized, laughed, and cried countless long and short hours with since the first second of this grand adventure, and who have outdone themselves month after month for almost a year now—I feel privileged to work with you in Berlin and everywhere. Thanks to the readers in Göttingen who picked me up from my desk evening after evening, to Robert, Stephan, Holger, and Annalena and the team in Berlin, who were not only there when needed, but also had my back during the long nights. To Kate, Nicolas, and the team at 350.org, from whom I was able to learn so much before we started storming the streets. Thanks to my dearest souls from Hamburg for unfailingly strengthening my nerves, and to those who have supported me for so many hours as a matter of course in the back office or with wine. And finally, thank you to the great visionaries who have influenced, inspired, and accompanied me in climate science—geographically, socially, culturally, politically, and philosophically.

ALEX A big thank you to my mother, who knows how to turn anger into wisdom, my siblings, my father, and my entire family for their unconditional support over the years. Thanks to the S12 for indulgence in turbulent times, to Hannes and Vincent, Christoph, Aimo, Lea, Anselm, Anna Sofie, and Valerie, for roots and wings in the past months; to Nico, Jonas, Marie, Patrick, Valentin, Hannah, Benni, Andreas, and Fiona. Thanks for inspiration, comments, and wise thoughts at all times of day and night. Thank you to the many role models at Cusanus University and in the Rhineland, to whom I owe so much, and to the trailblazers who opened my eyes with their commitment to the climate strikes, Climate Camps, and to *Ende Gelände*. And a big bow to the countless heroes and heroines who work invisibly behind the scenes, who do the care work, and who stand up against injustice and empower others to do the same.

SABINE Thanks to Sue Ramin from Brandeis University Press for supporting this publication, and to Alan Berolzheimer and especially Jim Schley for careful editing. Thanks also to my son Jonathan for his critical feedback and ready advice throughout. To him, his big brother David, and all their cousins I dedicate my contribution to this project. Special thanks to my dear friend Jennifer for her careful reading of the text and her many helpful suggestions, and to the BigWigs for their wisdom, friendship, and guidance over more than two decades. A shoutout to my brother Friedhelm for always being a step ahead, to Sabine, Ruth, Tine, Anne, Elke, and Laura for help with those tricky words, and to my parents for providing a safe haven at all times. My friends in 350Mass, NoCoalNoGas, Transition Wayland, and my fellow water protectors in the fight against Line3—above all Marla, who rocks out front and behind the scenes: You do what must be done. Thanks to my students at Brandeis University, whose commitment to charting a different course continuously encourages me to go the extra mile.

BRANDEIS UNIVERSITY PRESS acknowledges the generous support of the Theodore and Jane Norman Fund, which helped to produce this book. We also acknowledge with gratitude the generosity of the Center for German and European Studies at Brandeis University, which is supported by the German Academic Exchange Service (DAAD) with funds from the German Federal Foreign Office (Auswärtiges Amt.)

NOTES

PREFACE

1. https://grain.org/en/article/6634-corporate-greenwashing-net-zero-and-nature-based-solutions-are-a-deadly-fraud#sdfootnote1anc/.

2. https://redd-monitor.org/2019/04/09/shell-and-natural-climate-solutions-us300-million-for-carbon-offsets/.

INTRODUCTION

1. Bundesministerium für Ernährung und Landwirtschaft, *Welternährung verstehen. Fakten und Hintergründe* (Understanding World Nutrition. Facts and Background) (2018), www.bmel.de/SharedDocs/Downloads/DE/Broschueren/Welternaehrung-verstehen.html.

2. Welthungerhilfe, "Hunger, Verbreitung, Ursachen & Folgen" (Hunger: Prevalence, Causes, & Consequences) (2019), www.welthungerhilfe.de/hunger/.

3. UN Refugee Agency, "Refugee Figures" (2022) www.unhcr.org/figures-at-a-glance.html.

4. World Health Organization, "Depression: Let's Talk" (2019), www.who.int/mental_health/management/depression/en/.

5. Translator's note: By "crisis of social reproduction," the authors mean the reproduction of all structures and conditions that are necessary for human life, including paid and unpaid care work.

6. Ulrich Brand and Markus Wissen, *The Imperial Mode of Living: On the Exploitation of People and Nature in Global Capitalism* (Berkeley, CA: Verso, 2017).

7. "Klimawandel Konsequenzen des Klimawandels für Hamburg" (Climate Change Consequences for Hamburg), available at: www.hamburg.de/klimawandel-in-hamburg/.

8. Achim Daschkeit and Anna Luisa Renken, "Klimaänderung und Klimafolgen in Hamburg. Fachlicher Orientierungsrahmen" (Climate Change and Climate Consequences in Hamburg. Expert Orientation Framework) (2009), www.hamburg.de/contentblob/4434742/7c76e9c6509b9ca39cb4f8f3aff20805/data/d-orientierungsrahmen-klimawandel-in-hamburg.pdf.

9. Daschkeit and Renken, "Klimaänderung."

10. Florence Gaub, "Global Trends to 2030: Challenges and Choices for Europe" (2019), https://espas.secure.europarl.europa.eu/orbis/sites/default/files/generated/document/en/ESPAS_Report2019.pdf.

11. David Spratt and Ian Dunlop, "Existential Climate-Related Security Risk: A Scenario Approach" (2019), www.semanticscholar.org/paper/Existential-climate-related-security-risk%3A-a-Spratt-Dunlop/2b496862de471adbf44982470f5525b6950ea37c.

12. Harald Koisser, "Wir Possibilisten" (We Possibilists) (2017), www.koisser.at/wir-possibilisten/.

1 OUR FUTURE IS A DYSTOPIA

1. Tagesthemen, Sendung (broadcast), January 25, 2019, 9:10 p.m., www.youtube.com/watch?v=efzqwAfPsGU.

2. Climate Action Tracker, 2100 Warming Projections, December 11, 2018, https://climateactiontracker.org/global/temperatures/.

3. David Spratt, "What Would 3 Degrees Mean?" *Climate Code Red,* September 1, 2010, www.climatecodered.org/2010/09/what-would-3-degrees-mean.html/.

4. Intergovernmental Panel on Climate Change, "Special Report: Global Warming of 1.5°C: Summary for Policymakers" (2015), https://www.ipcc.ch/sr15/.

5. US Environmental Protection Agency, "Climate Impacts on Coastal Areas" (2016), https://19january2017snapshot.epa.gov/climate-impacts/climate-impacts-coastal-areas/.

6. Will Steffen, Johan Rockström, Katherine Richardson, Timothy Lenton, Carl Folke, and Diana Liverman, et al., "Trajectories of the Earth System in the Anthropocene," *Proceedings of the National Academy of Sciences of the United States of America,* 115, no. 33 (2018): 8252–59.

7. Translator's note: After World War II, especially in West Germany and Austria but also in other parts of Europe, the swift economic upswing in the 1950s and 1960s was dubbed an "economic miracle."

8. Arthur Sullivan, "Der Klimawandel und das Fliegen" (Climate Change and Flying), *Deutsche Welle,* January 10, 2018, www.dw.com/de/der-klimawandel- and-flying/a-42094220/.

9. Cornelia Koppetsch, "Generation Y," *Deutschlandfunk Nova,* October 13, 2018, www.deutschlandfunknova.de/beitrag/soziologie-generation-y/.

10. World Health Organization, "Diabetes Country Profiles, Nauru" (2016), www.who.int/publications/m/item/diabetes-nru-country-profile-nauru-2016.

11. Luc Folliet, *Nauru, die verwüstete Insel: Wie der Kapitalismus das reichste Land der Welt zerstörte* (Nauru: The Devastated Island: How Capitalism Destroyed the Richest Country on Earth) (Berlin: Wagenbach, 2011).

12. Amnesty International, "Nauru 2017/18" (2018), www.amnesty.org/en/location/asia-and-the-pacific/south-east-asia-and-the-pacific/nauru/report-nauru/.

13. Refugee Action Coalition Sydney, "Manus and Nauru," www.refugeeaction.org.au/?page_id=4528 (2019). See also www.theguardian.com/australia-news/2021/sep/24/australia-signs-deal-with-nauru-to-keep-asylum-seeker-detention-centre-open-indefinitely.

14. United Nations Development Programme, "Nauru" (2019), www.adaptation-undp.org/explore/micronesia/nauru.

2 BECAUSE YOU ARE STEALING OUR FUTURE

1. Bundesministerium für Umwelt, Naturschutz und nukleare Sicherheit, "Zukunft? Jugend fragen!" (Future? Ask Youth!) (2018), www.bmuv.de/fileadmin/Daten_BMU/Pools/Broschueren/jugendstudie_bf.pdf.

2. The term Anthropocene blinds us to the fact that it is not human beings per se who are altering the planetary systems, but the imperial way of living predominant in industrialized societies. Several scholars suggest the term "Capitalocene" instead, highlighting the role of capitalism in the changing global biological, geological, and atmospheric processes. See: Jason W. Moore, *Anthropocene or Capitalocene?: Nature, History, and the Crisis of Capitalism* (Oakland, CA: PM Press, 2016).

3. Intergovernmental Panel on Climate Change, "Special Report: Global Warming of 1.5°C: Summary for Policymakers" (2015), www.ipcc.ch/sr15/. Also see Philip Shabecoff, "Global Warming Has Begun, Expert Tells Senate," *New York Times*, June 24, 1988, www.nytimes.com/1988/06/24/us/global-warming-has-begun-expert-tells-senate.html/; and Stephan Rahmstorf, "Paläoklima: Das ganze Holozän" (Paleoclimate: The Whole Holocene), *Spektrum.de SciLogs*, June 17, 2013, https://scilogs.spektrum.de/klimalounge/palaeoklima-das-ganze-holozaen/.

4. The World Bank, "CO_2 Emissions (Metric Tons Per Capita)" (2019), https://data.worldbank.org/indicator/EN.ATM.CO2E.PC?contextual=max&end=2014&locations=US&start=1960. Also see, always accessible and up to date: https://scripps.ucsd.edu/programs/keelingcurve/.

5. Nathaniel Rich, *Losing Earth: A Recent History* (New York: MCD, 2019).

6. See www.economist.com/graphic-detail/2021/07/08/a-third-of-americans-deny-human-caused-climate-change-exists.

7. German Bundestag, "Erster Zwischenbericht der Enquete Kommission Vorsorge zum Schutz der Erdatmosphäre" (First Interim Report of the Enquete Commission. Precautionary Measures for the Protection of the Earth's Atmosphere). Printed matter 11/3246 (2.11.1988): 3.

8. United Nations Framework Convention on Climate Change, "Adoption of the

Paris Agreement" (2015), https://unfccc.int/ resource/docs/2015/cop21/eng /l09r01.pdf.

9. Deutscher Bundestag, *Das Grundgesetz.* (Basic Law=German Constitution) (2019), www.bundestag.de/grundgesetz.

10. Bundesministerium für Umwelt, "Zukunft?"

11. Though being an important milestone for global climate policy, bringing many more countries into the agreement, the Paris Agreement of 2015 left some loopholes that threaten to undermine real mitigation efforts. It saw a retreat from obligations of the earlier Kyoto Protocol in favor of voluntary national contributions. And it outlined a mechanism for global carbon trading, whose dangers we discuss in more detail in chapter 7.

12. See for example www.smithsonianmag.com/smart-news/europes-extreme -floods-nine-times-more-likely-because-climate-change-180978519/.

13. Even this calculation does not give us a 100 percent chance to meet the 1.5 degrees Celsius. Emitting a maximum of 308 gigatons is expected to give us a 66 percent chance, according to the Mercator Research Institute on Global Commons and Climate Change.

14. Mercator Research Institute on Global Commons and Climate Change, "That's How Fast the Carbon Clock is Ticking" (2019), www.mcc-berlin.net/en /research/co2-budget.html.

15. Translator's note: CDU (Christlich Demokratische Union/Christian Democratic Union) is the main center-right party in Germany. Angela Merkel (CDU) was German Chancellor from 2005 until 2021. In September 2021, Olaf Scholz of the Social Democrats (SPD) became her successor.

16. Translator's note: Merkel herself had used the expression "No More *Pillepalle*" (Small Steps) in June 2019 to urge her own government to adopt more aggressive action on climate.

17. Translator's note: The *Kirchentag* is a large biannual gathering of German Protestants, organized by lay people, known for its open political debates of societal issues of the day. Barack Obama and Angela Merkel talked about democracy at the 2017 *Kirchentag*, for example.

18. In first place is China, whose companies are obviously state owned. This complicates this exemplary calculation, which we have overlooked in this case. Incidentally, Chinese coal production accounts for 14.3 percent of global greenhouse gases according to this report: Paul Griffin, "The Carbon Majors Database: Methodology Report 2017," https://cdn.cdp.net/cdp-production/comfy/cms/files/files/000 /000/979/original/Carbon-Majors-Database-2017-Method.pdf, (2017): 14.

3 WE LACK A UTOPIA

1. Jürgen Habermas, "Die Krise des Wohlfahrstaates und die Erschöpfung utopischer Energien" (The Crisis of the Welfare State and the Creation of Utopian Energies), in: Jürgen Habermas, *Die Moderne—Ein unvollendetes Projekt* (Modernism—An Unfinished Project), (Leipzig: Reclam Verlag, 1984), 105–29.

2. We speak of fossil capitalism because the energy generation, production, and infrastructure of our economy has been largely dependent on the use of fossil energy carriers such as coal, oil, and gas since the industrial revolution. These dependencies continue today. Currently, more than half of the world's ten largest companies by revenue make their money from this dirty energy source (for more details, see chapters 7 and 8).

3. Jean-François Lyotard, *The Postmodern Condition: A Report on Knowledge* (Minneapolis: University of Minnesota Press, 1984).

4. Francis Fukuyama, *The End of History* (New York: Free Press, 1992).

5. Mengpin Ge, Johannes Friedrich, and Thomas Damassa, "Six Graphs Explain the World's Top 10 Emitters," World Resources Institute, www.wri.org/blog/2014/11/6-graphs-explain-world-s-top-10-emitters (November 25, 2014). For more details, see chapter 4.

6. At this point, the time CO_2 stays in the atmosphere is a significant factor, but also a rather complicated one. A good overview is provided, for example, by this translation of the FAQ from a 2014 IPCC report: "Climate FAQ 12.3 | Emissions," German Climate Consortium (2019), www.deutsches-klima-konsortium.de/en/klimafaq-12-3.html/.

7. Ian Tiseo, "Cumulative CO_2 emissions from 1750 to 2020," *Statista*, January 4, 2022, www.statista.com/statistics/1007454/cumulative-co2-emissions-worldwide-by-country/.

8. Marcia Rocha, Mario Krapp, Johannes Guetschow, M. Louise Jeffery, Bill Hare, and Michiel Schaeffer, "Historical Responsibility for Climate Change—From Countries Emissions to Contribution to Temperature Increase, Climate Analytics" (Potsdam Institute for Climate Impact Research, 2015), https://climateanalytics.org/media/historical_responsibility_report_nov_2015.pdf.

9. Since we published the first edition of this book in 2019, the discourse around "net-zero" targets has widely been hijacked for corporate greenwashing. Major corporations like Shell and Nestlé pledge to "offset" their emissions instead of changing their emission-heavy business models. See: https://grain.org/en/article/6634-corporate-greenwashing-net-zero-and-nature-based-solutions-are-a-deadly-fraud/.

10. The term "Global South" is not only a geographic designation, but also an analytical one. According to the organization glokal e.V., it describes "a social, political, and economic position that is disadvantaged in the global system" and replaces the

division of the world into developed and developing countries. See: www.glokal.org/publikationen/mit-kolonialen-gruessen/.

11. René Bocksch, "Die Welt ist nicht genug" (The World is Not Enough), March 24, 2022, *Statista*, https://de.statista.com/infografik/10574/oekologischer-fussabdruck-die-welt-ist-nicht-genug/.

12. The World Bank, "Total Greenhouse Gas Emissions (% Change from 1990)" (2019), https://data.worldbank.org/indicator/EN.ATM.GHGT.ZG/.

13. Claire Berlinski, *There Is No Alternative: Why Margaret Thatcher Matters* (New York: Basic Books, 2008).

14. Translator's note: Since 1991, a German civil society organization declares a word to be the most inappropriate word of the year in order to sensitize the public for the use of language. Words are chosen on the basis of criteria that include violation of human rights and democracy, and discrimination against minorities (www.unwortdesjahres.net/).

15. "Alternativlos ist das Unwort des Jahres" ("Without Alternative" Is the Worst Word of the Year), *Spiegel Online*, January 18, 2011, www.spiegel.de/kultur/gesellschaft/sprachkritik-alternativlos-ist-das-unwort-des-jahres-a-740096.html.

16. Silja Graupe and Harald Schwaetzer, "Bildungsorte transformativ-reflexiver Ökonomie" (Educational Sites of Transformative-Reflexive Economy), in: Reinhard Pfriem, Uwe Schneidewind, Johannes Barth, Silja Graupe, and Thomas Korbun, eds., *Transformative Wirtschaftswissenschaft im Kontext nachhaltiger Entwicklung* (Transformative Economics in the Context of Sustainable Development) (Marburg: Metropolis-Verlag, 2017), 509.

17. "Japan Commemorates Victims of Fukushima," *Deutsche Welle*, March 11, 2019, www.dw.com/de/japan-gedenkt-der-Opfer-von-Fukushima/a-47847882/.

18. Graupe and Schwaetzer, "Bildungsorte," 524ff.

19. Ibid., 526.

20. Translator's note: Armin Laschet was the former designated successor for Angela Merkel as CDU candidate for chancellor; he lost to Olaf Scholz (SPD) in September 2021.

21. Translator's note: Anne Will is a German talk show host.

22. "Verkehrsminister: Wir brauchen den Elektro-Käfer-Effekt" (Minister of Transportation: We Need the Electro-Beetle-Effect), *Süddeutsche Zeitung*, January 20, 2019, www.sueddeutsche.de/wirtschaft/auto-verkehrsminister-wir-brauchen-den-elektro-kaefer-effekt-dpa.urn-newsml-dpa-com-20090101-190130-99-788679.

23. Melanie Amann and Gerald Traufetter, "Meine Generation wurde betrogen" (My Generation was Cheated), *Spiegel Online*, March 15, 2019, www.spiegel.de/plus/luisa-neubauer-and-peter-altmaier-in-controversy-talk-a-00000000-0002-0001-0000-000162913137.

24. Uwe Jäger, "Lust auf Zukunft mit Harald Welzer" (Desiring a Future with

Harald Welzer), *Saarlandwelle*, March 27, 2019, www.ardaudiothek.de/episode/sr-3-aus-dem-leben/lust-auf-zukunft-mit-harald-welzer/sr-3-saarlandwelle/61509918/.

25. Robert Misik, "Hey, psst—ham Sie mal 'n Narrativ für die Linke übrig?" (Hey, Psst—Have a Narrative for the Left to Spare?) *taz*, January 14, 2017, https://taz.de/!5371521/.

4 THE CLIMATE CRISIS IS NOT AN INDIVIDUAL CRISIS

1. Garrett Hardin, "The Tragedy of the Commons," *Science*, 162, no. 3859 (1968): 1243–48.

2. In this context we are talking about CO_2, which is of course not the only greenhouse gas, but the most dangerous one.

3. Only the CO_2 that ends up in the atmosphere drives the greenhouse effect. Accordingly, it is desirable that other sinks store as much CO_2 as possible (which is why the protection of forests, for example, plays such an important role for the climate).

4. Volker Stollorz, "Elinor Ostrom und die Wiederentdeckung der Allmende. Aus Politik und Zeitgeschichte" (Elinor Ostrom and the Rediscovery of the Commons: From Politics and Contemporary History) (2011), www.bpb.de/ apuz/33204/elinor-ostrom-and-the-re-discovery-of-the-everything?p=all.

5. Susanne Preuss, "Bosch will bis 2020 klimaneutral sein" (Bosch Will Be Climate Neutral by 2020), *Frankfurter Allgemeine Zeitung*, May 9, 2019, www.faz.net/aktuell/wirtschaft/unternehmen/co2-bosch-will-ab-2020-komplett-klimaneutral-sein-16178383.html/.

6. Alexander Richter, "Costa Rica Reports Near 100% Renewable Energy Electricity Supply and Electricity Export," *Think Geo Energy*, August 4, 2019, www.thinkgeoenergy.com/costa-rica-reports-near-100-renewable-energy-electricity-supply-and-electricity-export/.

7. Michael Kopatz, Ökoroutine. Damit wir tun, was wir für richtig halten (Ecoroutine. So We Do What We Think Is Right) (Munich: Oekom Verlag, 2016).

8. DW Culture (@DWCulture), "We dug deep into our archives and found this absolute gem: an interview with #AngelaMerkel from the year 1995. These days the whole world knows her name; back then the host seemed to still be having trouble with it. #merkel#interview #archive #1995," Twitter, July 16, 2019, 8:21a.m., https://twitter.com/dw_culture/status/1151104545357008896?s=20/.

5 THE CLIMATE CRISIS IS A CRISIS OF RESPONSIBILITY

1. Nathaniel Rich, "Losing Earth: The Decade We Almost Stopped Climate Change," *New York Times*, August 1, 2018, www.nytimes.com/interactive/2018/08/01/magazine/climate-change-losing-earth.html.

2. Translator's note: Article 20a of the German Constitution, www.gesetze-im-internet.de/englisch_gg/englisch_gg.html#p0116.

3. Deutscher Bundestag, "Wie Umwelt- und Tierschutz ins Grundgesetz kamen" (How the Environment and Animal Welfare Entered into the Basic Law) (2013), www.bundestag.de/dokumente/textarchiv/2013/47447610_kw49_grundgesetz_20a-213840/. See also note 11, below.

4. SEJM, *Constitution of the Republic of Poland* (1997), available at: www.sejm.gov.pl/prawo/konst/angielski/kon1.htm/.

5. Thuringia State Parliament, *Constitution of the Free State of Thuringia* (2010), www.thueringer-landtag.de/fileadmin/user_upload/Verf_TH.pdf, 5.

6. Hans Jonas, *Das Prinzip Verantwortung. Versuch einer Ethik für die technologische Zivilisation* (The Principle of Responsibility: Attempt at an Ethics for the Technological Civilization) (Frankfurt am Main: Suhrkamp Taschenbuch Verlag, 1984), 70. See also REVOSax, *Constitution of the Free State of Saxony* (2014), www.revosax.sachsen.de/vorschrift/3975-Saechsische-Constitution#a10/.

7. Jonas, *Das Prinzip Verantwortung,* 70.

8. Ibid., 7.

9. Ibid., 174.

10. Ibid., 234.

11. A few months after this book was published in German, Luisa and a number of other young people sued the German government over its climate inaction. The case was called *Neubauer vs Government.* In a historic decision, they won. On April 29, 2021, the German Constitutional Court partially struck down the government's 2019 climate law, stating that "the fundamental rights—as intertemporal guarantees of freedom—afford protection against the greenhouse gas reduction burdens imposed by Article 20a of the Basic Law being unilaterally offloaded onto the future." See: www.bundesverfassungsgericht.de/SharedDocs/Pressemitteilungen/EN/2021/bvg21-031.html.

12. Greenpeace.de, "Regierung zur Rechenschaft" (Government to Account), by Michael Weiland, posted October 27, 2018, www.greenpeace.de/klimaschutz/klimakrise/klimaklage-anwaeltin-roda-verheyen-interview.

13. "Niederlande werden zu Klimaschutz gezwungen" (Netherlands Forced to Protect Climate), *Frankfurter Allgemeine Zeitung,* October 9, 2018, www.faz.net/aktuell/wirtschaft/gerichtsurteil-niederlande-werden-zu-klimaschutz-gezwungen-15829057/die-umweltschuetzer-von-urgenda-15829124.html.

14. "Letzte Instanz für den Klimaschutz: Klimaklagen weltweit" (Court of Last Resort for Climate Action: Climate Lawsuits Worldwide), Greenpeace.de, 2018, 2, www.greenpeace.de/publikationen/factsheet-klimaklagen-weltweit. See also https://en.klimaseniorinnen.ch/.

15. Ibid., 3.

16. Ibid., 4.

17.Ibid., 5.

18. Günther Anders, *Die atomare Drohung. Radikale Überlegungen zum Atomzeitalter* (The Nuclear Threat: Radical Reflections on the Nuclear Age) (München: C. H. Beck, 1993), 1ff.

19. Curt D. Storlazzi, Stephen B. Gingerich, Ap Van Dongeren, Olivia M. Cheriton, Peter W. Swarzenski, and Ellen Quataert, et al., "Most Atolls Will Be Uninhabitable by the Mid-21st Century Because of Sea-level Rise Exacerbating Wave-Driven Flooding," *Science Advances,* 4 no. 4 (2018): 1–10.

20. "Greta Thunberg Speaks in Katowice: 'Our Leaders Behave Like Children,'" *Dagens Nyheter,* March 12, 2018, www.dn.se/kultur-noje/greta-thunberg-speaks-in-katowice-our-leaders-behave-like-children/.

21. Jonas, *Das Prinzip Verantwortung,* 36.

6 THE CLIMATE CRISIS IS A CRISIS OF COMMUNICATION

1. Translator's note: *Friedrichstadtpalast* is an entertainment venue in Berlin.

2. NASA, "Sea level" (2019), https://climate.nasa.gov/vital-signs/sea-level/.

3. C40 Cities, "Sea Level Rise and Coastal Flooding" (2022), www.c40.org/what-we-do/scaling-up-climate-action/adaptation-water/the-future-we-dont-want/sea-level-rise/.

4. See https://climate.mit.edu/ask-mit/why-do-some-people-call-climate-change-existential-threat/.

5. Elisabeth Wehling, *Politisches Framing: wie eine Nation sich ihr Denken einredet—und daraus Politik macht* (Politics and Framing: How a Nation Convinces Itself of Its Thinking—and Turns That into Politics) (Köln: Herbert von Halem Verlag, 2016), 21.

6. Ibid., 30.

7. George Lakoff, "Why It Matters How We Frame the Environment," *Environmental Communication,* 4, no. 1 (2010), 70–80.

8. Ibid., 42.

9. Ibid., 185.

10. Graham Readfearn, "Doubt Over Climate Science is a Product with an Industry Behind It," *Guardian,* March 5, 2015, www.theguardian.com/environment/planet-oz/2015/mar/05/doubt-over-climate-science-is-a-product-with-an-industry-behind-it/.

11. Naomi Oreskes, "The Scientific Consensus on Climate Change," *Science* 306, no. 5702 (2004): 1686.

12. Maxwell T. Boykoff and Jules M. Boykoff, "Balance as Bias: Global Warming and the US Prestige Press," *Global Environmental Change* 14, no. 2 (2004): 125–36.

13. Naomi Oreskes and Eric M. Conway, *Merchants of Doubt: How a Handful of Scientists Obscured Truth on Issues from Tobacco Smoke to Climate Change* (London: Bloomsbury Publishing, 2011).

14. Ibid., xxi.

15. Sandra Laville, “Top Oil Firms Spending Millions Lobbying to Block Climate Change Policies, Says Report,” *Guardian*, March 22, 2019, www.theguardian.com/business/2019/mar/22/top-oil-firms-spending-millions-lobbying-to-block-climate-change-policies-says-report/.

16. Initiative Neue Soziale Marktwirtschaft, “12 Fakten zur Klimapolitik. Fortschritt, Wachstum und Klimaschutz gehörenzusammen” (12 Facts about Climate Policy: Progress, Growth, and Climate Protection Belong Together), June 27, 2019, www.insm.de/insm/kampagne/klimaschutz/12-fakten-zur-klimapolitik/.
For a fact check of the claims of Initiative Neue Soziale Marktwirtschaft, see Volker Quaschning, *Faktencheck der 12 Fakten zum Klimaschutz der Initiative Neue Soziale Marktwirtschaft* INSM (Fact Check of the 12 Facts About Climate Protection Initiative of Neue Soziale Marktwirtschaft INSM), July 17, 2019, www.volker-quaschning.de/artikel/Fakten-INSM/index.php/.

17. “Scientists Shocked by Arctic Permafrost Thawing 70 Years Sooner Than Predicted,” *Guardian*, June 18, 2019, https://www.theguardian.com/environment/2019/jun/18/arctic-permafrost-canada-science-climate-crisis/.

18. Tomasz Konicz, “Weltklima auf der Kippe” (World Climate on the Brink), *Telepolis*, July 3, 2019, www.heise.de/tp/features/Weltklima-auf-der-Kippe-4456028.html?seite=all/.

19. Alexander M. Petersen, Emmanuel M. Vincent, and Anthony L. Westerling, “Discrepancy in Scientific Authority and Media Visibility of Climate Change Scientists and Contrarians,” *Nature Communications* 10, no. 1 (2019): 3502.

20. Günther Anders, *Die atomare Drohung. Radikale Überlegungen zum Atomzeitalter* (The Nuclear Threat: Radical Reflections on the Nuclear Age) (München: C. H. Beck, 1993), 98.

7 THE CLIMATE CRISIS IS A CRISIS OF FOSSIL CAPITALISM

1. Johan Rockström, Will Steffen, Kevin Noone, Åsa Persson, F. Stuart Chapin III, and Eric Lambin, et al., “Planetary Boundaries: Exploring the Safe Operating Space for Humanity,” *Ecology and Society* 14 (2009): 1–33.

2. Christian Schwägerl, “Dramatischer Uno-Bericht. Eine Million Arten vom Aussterben bedroht” (Dramatic UN Report: One Million Species Threatened by Extinction), *Spiegel Online*, May 6, 2019, www.spiegel.de/wissenschaft/natur/artensterben-uno-bericht-beschreibt-dramatischen-verlust-der-artenvielfalt-a-1265482.html/.

3. "Bei Klimapolitik. Merkel will nicht über Verbote und Gebote arbeiten" (In Climate Policy: Merkel Does Not Want to "Work with Bans and Regulations"), *Frankfurter Allgemeine Zeitung*, August 22, 2018, www.faz.net/-gpg-9qd3b/.

4. Friedrich A. Hayek, *The Fatal Conceit: The Errors of Socialism* (Chicago: University of Chicago Press, 1988).

5. Stephan Pühringer, "Marktmetaphoriken in Krisennarrativen von Angela Merkel" (Market Metaphorics in Crisis Narratives by Angela Merkel), in Walter Ötsch, Katrin Hirte, Stephan Pühringer, and Lars Bräutigam, eds., *Markt! Welcher Markt? Der interdisziplinäre Diskurs um Märkte und Marktwirtschaft* (Market! Which Market? The Interdisciplinary Discourse about Markets and Market Economy) (Marburg: Metropolis-Verlag, 2015), 229–51.

6. Walter Ötsch, *Mythos Markt, Mythos Neoklassik. Das Elend des Marktfundamentalismus* (Myth of the Market, Myth of Neoclassicism: The Misery of Market Fundamentalism) (Marburg: Metropolis-Verlag, 2019), 39ff.

7. Ibid., 29.

8. Translator's note: *Hart aber Fair* ("Tough But Fair") is a talk show on German public television.

9. See https://icapcarbonaction.com/en/ets-map/.

10. The World Bank, "State and Trends of Carbon Pricing 2021," May 2021 (World Bank, Washington, DC). Doi: 10.1596/978-1-4648-1728-1. License: Creative Commons Attribution CC BY 3.0 IGO.

11. S. Fiedler and K. Geppert, "Empirische Analysen zum Emissionshandel" (Empirical Analyses of Emissions Trading) *DIWWochenbericht* 9, no. 83 (2016):170–84, see 172.

12. Karl Polanyi, *The Great Transformation: Political and Economic Origins of Our Time*(Boston: Beacon Press, 1944), xxv.

13. Ibid., 243.

14. Ibid., 244.

15. In Roger Revelle and Hans E. Suess, "Carbon Dioxide Exchange Between Atmosphere and Ocean and the Question of an Increase of Atmospheric CO_2 Duringthe Past Decades," *Tellus* 9, no. 1 (1957): 19.

16. Intergovernmental Panel on Climate Change, "Special Report: Global Warming of 1.5°C Mitigation Pathways Compatible with 1.5°C in the Context of Sustainable Development" (2015), available at www.ipcc.ch/sr15/chapter /chapter-2/.

17. Ibid., available at www.ipcc.ch/sr15/chapter/chapter-5/.

18. Karl Marx and Friedrich Engels, *The Eighteenth Brumaire of Louis Bonaparte* (1852; London: The Electric Book Company, 2001), 7.

19. The EEG (Erneuerbare Energien Gesetz), or Renewable Energy Law, was introduced by the Red/Green Coalition in 2000.

8 THE CLIMATE CRISIS IS A CRISIS OF PROSPERITY

1. Translator's note: While the term "shitstorm" is not new, it is now most frequently used when a person gets attacked through a wave of vociferous posts, comments, etc., on social media.

2. Hans Rosling, Ola Rosling, and Anna Rosling Rönnlund, *Factfulness* (London: Sceptre, 2018), 3–5.

3. Bruno Urmersbach, "Weltweites Bruttoinlandsprodukt (BIP) bis 2018" (Global GDP until 2018), *Statista,* August 9, 2019, https://de.statista.com/statistik/daten/studie/159798/umfrage/entwicklung-des-bip-bruttoinlandsprodukt-weltweit/.

4. Rosling, et al., *Factfulness,* 13.

5. Ibid., 47.

6. Ibid., 135.

7. "10 Biggest Corporations Make More Money Than Most Countries in the World Combined," *Global Justice Now,* September 12, 2016, www.globaljustice.org.uk/news/2016/sep/12/10-biggest-corporations-make-more-money-most-countries-world-combined/.

8. Kate Raworth, *Doughnut Economics: Seven Ways to Think Like a 21st-Century Economist* (London: Chelsea Green Publishing, 2017), 13.

9. Martin Kolmar, "Immer mehr Wachstum wird unser Leben zerstören" (More and More Growth Will Destroy Our Lives), *Zeit Online,* June 14, 2019, www.zeit.de/wirtschaft/2019-04/industriepolitik-umstieg-klimapolitik-digitalisierung-globalisierung-nachhaltigkeit/.

10. Hans Böckler Stiftung, "Wohlstand in Deutschland wegen erhöhter Ungleichheit nur auf Niveau der 1990er Jahre—2016 erneut leichte Verbesserung" (Prosperity in Germany Due to Heightened Inequality Only on the Level of the 1990s—Slight Improvement Again in 2016), July 19, 2018, www.boeckler.de/de/presse mitteilungen-15992-wohlstand-in-deutschland-wegen-erhoehter-ungleichheit-nur-auf-niveau-der-1990er-jahre-3147.htm/.

11. "69 of the Richest 100 Entities on the Planet are Corporations, not Governments, Figures Show," *Global Justice Now,* October 17, 2018, www.globaljustice.org.uk/news/69-richest-100-entities-planet-are-corporations-not-governments-figures-show/.

12. Facundo Alvaredo, Lucas Chancel, Thomas Piketty, Emmanuel Saez, and Gabriel Zucman, "World Inequality Report 2018," *World Inequality Lab* (2018), https://wir2018.wid.world/files/download/wir2018-summary-english.pdf.

13. Ulrich Brand and Marcus Wissen, *The Imperial Mode of Living: Everyday Life and the Ecological Crisis of Capitalism* (Berkeley, CA: Verso, 2021).

14. Global Carbon Atlas, "Fossil Fuel Emissions" (2018), www.globalcarbonatlas

.org/en/CO2-emissions/. Even through the disruptions of the global pandemic, CO_2 emissions continued to rise, leaving CO2 in the atmosphere at an all-time high of 418 ppm in early 2022. See: https://climate.nasa.gov/vital-signs/carbon-dioxide/.

15. Luise Tremel, "Aufhören. Warum, wie, wer und wann am Besten was" (To Stop: Why, How, Who, and When What), January 26, 2017, https://netzwerk-n.org/blog/aufhoeren-warum-wie-wer-und-wann-am-besten-was/.

16. Ibid., 5.

17. Raworth, *Doughnut Economics,* 10.

18. Ibid., 25.

19. Ibid., 28.

20. Ibid.

21. Ibid., 29.

22. Ibid.

23. Ibid., 30.

24. Thomas Fatheuer, *Buen Vivir: Eine kurze Einführung in Lateinamerikas neue Konzepte zum guten Leben und zu den Rechten der Natur* (Buen Vivir: A Short Introduction to the New Latin-American Concept of the Good Life and the Rights of Nature) (Berlin: Heinrich-Böll-Stiftung, 2011), 16.

25. Peter Carstens, "Maori-Fluss erhält Rechte als Person" (Maori River Receives Personhood Rights), *GEO,* March 13, 2017, www.geo.de/natur/nachhaltigkeit/15997-rtkl-neuseeland-maori-fluss-erhaelt-rechte-als-person/.

26. Markus Söder: chair of the Christlich-Soziale Union (Christian Social Union, CSU), the CDU's sister party in Bavaria.

27. Initiative Neue Soziale Marktwirtschaft, "Nachhaltigkeit als Verfassungsprinzip. Eine bessere Politik für morgen mit Nachhaltigkeit im Grundgesetz" (Sustainability as a Constitutional Principle: A Better Politics for Tomorrow with Sustainability in the Basic Law), May 22, 2019, www.insm.de/insm/publikationen/studien/praesident-des-bundesverfassungsgerichts-ad-nachhaltigkeit-ins-grundgesetz/.

28. World Future Council, "Future Policy Award" (2019), www.worldfuturecouncil.org/de/future-policy-award/.

9 THE CLIMATE CRISIS IS A CRISIS OF JUSTICE

1. Barbara Hans, "Hurrikan Katrina. Wie aus New Orleans eine Stadt der Weißen wurde" (Hurricane Katrina: How New Orleans Became a City of White Folk), *Spiegel Online,* August 29, 2008, www.spiegel.de/panorama/gesellschaft/hurrikan-katrina-wie-aus-new-orleans-eine-stadt-der-weissen-wurde-a-575012-2.html/.

2. Dagmar Dehmer, "Zehn Jahre nach Hurrikan Katrina New Orleans—eine

gespaltene Stadt" (Ten Years After Hurricane Katrina New Orleans—A Divided City), *Tagesspiegel,* August 29, 2015, www.tagesspiegel.de/gesellschaft/panorama/new-orleans--eine-gespaltene-stadt-5192566.html/.

3. Damian Carrington, "Climate Apartheid: UN Expert says Human Rights May Not Survive," *Guardian,* June 25, 2019, www.theguardian.com/environment/2019/jun/25/climate-apartheid-united-nations-expert-says-human-rights-may-not-survive-crisis/.

4. FDP, "Lindner—Rede auf dem 69.FDP-Bundesparteitag" (Lindner—Speech at the 69th FDP Party Convention), May 12, 2018, available at https://mailings.fdp.de/node/123621/.

5. Schweizerische Eidgenossenschaft, "1987: BrundtlandReport" (2013), www.are.admin.ch/are/en/home/media/publications/sustainable-development/brundtland-report.html/.

6. Ende Gelände press release: "'Ende Gelände' Sets Out for Mass Action of Civil Disobedience + Activists Call for Climate Justice and Exit from Fossil Fuels in Buir," November 5, 2017, www.ende-gelaende.org/en/press-release/press-release-november-5th-2017-1111-a-m/.

7. Welt-sichten, "Klimawandel und Armut. Eine Herausforderung für gerechte Weltpolitik" (Climate Change and Poverty: A Challenge for a Just Global Politics) (2008), www.welt-sichten.org/dossiers/29368/klimawandel-und-armut/.

8. Carrington, "Climate Apartheid."

9. Tilman Santarius, "Emissionshandel und globale Gerechtigkeit" (Emissions Trading and Global Justice), in Susanne Böhler, Daniel Bongardt, and Siegfried Frech, eds., *Jahrhundertproblem Klimawandel. Forschungsstand, Perspektiven, Lösungswege* (Problem of the Century, Climate Change: State of Research, Perspectives, Paths to Solutions) (Schwalbach: Wochenschau Verlag, 2009), 121–38, www.santarius.de/645/emissionshandel-und-globale-gerechtigkeit/.

10. Intergovernmental Panel on Climate Change, "Special Report: Global Warming of 1.5°C: Sustainable Development, Poverty Eradication, and Reducing Inequalities" (2015), www.ipcc.ch/sr15/chapter/chapter-5/, Section 5.5.2.

11. A. Breitkopf, "ProKopfCO_2Emissionen nach ausgewählten Ländern weltweit im Jahr 2016 (in Tonnen)" (Per Capita CO_2 Emissions According to Select Countries Worldwide in 2016 [in Tons]), *Statista* (2019), https://de.statista.com/statistik/daten/studie/167877/umfrage/co-emissionen-nach-laendern-je-einwohner/.

12. Daniel W. O'Neill, Andrew L. Fanning, William F. Lamb, and Julia K. Steinberger, "A Good Life for All Within Planetary Boundaries," *Nature Sustainability* 1, no. 2 (2018): 88–95.

13. Our World in Data, "CO_2 Emissions per Capita, 2017: Average Carbon Dioxide (CO_2) Emissions Per Capita Measured in Tons Per Year" (2019), https://ourworldindata.org/grapher/co-emissions-per-capita/.

14. Rockström said this, among other things, during a speech at the Evangelical Church Congress 2019, where I, Luisa, was also present as a guest speaker.

15. Zoe Tabary, "Climate Change a 'Man-made Problemwith a Feminist Solution' Says Robinson," *Reuters*, June 18, 2018, www.reuters.com/article/us-global-climatechange-women/climate-change-a-man-made-problem-with-a-feminist-solution-says-robinson-idUSKBN1JE2IN/.

16. Oxfam, "An Economy that Works for Women: Achieving Women's Economic Empowerment in an Increasingly Unequal World," *Oxfam Briefing Paper* (2017), https://policy-practice.oxfam.org/resources/an-economy-that-works-for-women-achieving-womens-economic-empowerment-in-an-inc-620195/.

17. Deloitte, "Women in the Boardroom: A Global Perspective" (2017), www2.deloitte.com/global/en/pages/risk/cyber-strategic-risk/articles/women-in-the-boardroom-global-perspective.html/.

18. World Health Organization, "Gender, Climate Change, and Health" (2014), https://apps.who.int/iris/bitstream/handle/10665/144781/9789241508186_eng.pdf?sequence=1&isAllowed=y/.

19. United Nations Framework Convention on Climate Change, "Adoption of the Paris Agreement" (2015), https://unfccc.int/resource/docs/2015/cop21/eng/l09r01.pdf.

20. United Nations Framework Convention on Climate Change, "Introduction to Gender and Climate Change" (2019), https://unfccc.int/gender/.

21. Right Livelihood Award, "Medha Patkar and Baba Amte/Narmada Bachao Andolan" (2016), https://rightlivelihood.org/the-change-makers/find-a-laureate/medha-patkar-and-baba-amte-narmada-bachao-andolan/.

22. Translator's note: "The black-and-yellow coalition" was a German governing alliance during the years 1982–1998, made up of CDU/CSU (black) and FDP (yellow).

23. Chris Vielhaus, "Du zahlst zu viele Steuern. Aber aus anderen Gründen, als du denkst" (You Pay Too Much in Taxes. But for Different Reasons than You Think), *Perspective Daily*, April 30, 2019, https://perspective-daily.de/article/792/.

24. Stefan Bach and Andreas Thiemann, "Vermögensteuer" (Wealth Tax), *DIW Wochenbericht* 4 (2016): 79–89.

25. See www.oxfam.org/en/press-releases/over-100-millionaires-call-wealth-taxes-richest-raise-revenue-could-lift-billions/.

26. Arthur Sullivan, "Der Klimawandel und das Fliegen" (Climate Change and Flying), *Deutsche Welle*, January 10, 2018, www.dw.com/de/der-klimawandel-und-das-fliegen/a-42094220/.

27. "Flugreisen: Deutsche fliegen so viel wie nie" (Air Travel: Germans Fly Like Never Before), *Zeit Online*, January 17, 2019, www.zeit.de/mobilitaet/2019-01/flugreisen-luftverkehr-passagierrekord-flughafen-reiseaufkommen-anstieg/.

28. John Grant and Keith Baker, "How Will We Travel the World in 2050?" *The Conversation*, January 19, 2019, https://theconversation.com/how-will-we-travel-the-world-in-2050-121713/.

29. Reinhard Wollf, "Auf der Schiene oder gar nicht" (On Track or Not at All), *Klimareporter.de*, November 14, 2018, www.klimareporter.de/verkehr/auf-der-schiene-oder-gar-nicht/.

10 EDUCATE YOURSELVES!

1. Toralf Staud, "Problembewusstsein für den Klimawandel ist groß in Deutschland—am größten unter Frauen und im Westen" (Awareness of Climate Change is High in Germany—Highest among Women and in the West), *Klimafakten.de*, January 19, 2018, www.klimafakten.de/meldung/problembewusstsein-fuer-den-klimawandel-ist-gross-deutschland-am-groessten-unter-frauen-und/.

2. Joe Myers and Kate Whiting, "These Are the Biggest Risks Facing Our World in 2019," *World Economic Forum*, January 16, 2019, www.weforum.org/agenda/2019/01/these-are-the-biggest- risks-facing-our-world-in-2019/.

3. BlackRock, "Investors Underappreciate Climate-Related Risks in Their Portfolios—BlackRock Report," April 4, 2019, www.blackrock.com/corporate/newsroom/press-releases/article/corporate-one/press-releases/investors-under appreciate-climate-related-risks-in-their-portfolios/.

4. Laurie Goering, "Without Climate Action, Economic Growth Will Be Reversed, Economist Warns," *Reuters*, July 3, 2019, www.reuters.com/article/us-climate-change-britain/without-climate-action-economic-growth-will-be-reversed-economist-warns-idUSKCN1TY2RW/.

5. https://ec.europa.eu/clima/eu-action/european-green-deal_en/.

6. V. Pawlik, "Anzahl der Personen in Deutschland, die sich selbst als Vegetarier einordnen oder als Leute, die weitgehend auf Fleisch verzichten, von 2007 bis 2021" (Number of Vegetarians in Germany, 2007–2021), *Statista*, August 9, 2019, https://de.statista.com/statistik/daten/studie/173636/umfrage/lebenseinstellung-anzahl-vegetarier/.

7. Annett Entzian, "Denn sie tun nicht, was sie wissen" (For They Do Not Do What They Know) Ökologisches *Wirtschaften* 4 (2016): 21–23.

8. Friederike Otto, *Wütendes Wetter: Auf der Suche nach den Schuldigen für Hitzewellen, Hochwasser und Stürme* (Angry Weather: The Search for Those Responsible for Heat Waves and Storms) (Berlin: Ullstein, 2019).

9. The court has agreed to take on the case and judges are planning to visit the lake in Peru with court-appointed experts in summer 2022 (www.germanwatch.org/en/85108). A 2021 study by Oxford University scientists confirmed that human emissions are responsible for the warming that leads to the glacier retreat

above Huaraz. See: https://stiftungzukunft.org/en/study-supports-climate-litigation-claim-against-rwe-human-made-emissions-are-responsible-for-glacial-flood-risk-in-the-andes/.

10. Otto, *Wütendes Wetter*. See also: www.climate.gov/news-features/blogs/beyond-data/2021-us-billion-dollar-weather-and-climate-disasters-historical/.

11. Project Drawdown, *Solutions* (2019), www.drawdown.org/solutions/.

11 START DREAMING!

1. Günther Anders, *Die Antiquiertheit des Menschen. Über die Seelem Zeitalter der zweiten industriellen Revolution* (The Outdatedness of Man: About the Soul in the Age of the Second Industrial Revolution) (München: C. H. Beck, 1968), 273.

2. David Wallace-Wells, *The Uninhabitable Earth: Life After Warming* (New York: Tim Duggan Books, 2019).

3. Kathy Jetñil-Kijiner and Aka Niviâna, "Rise: From One Island to Another," 350.org (2018), https://350.org/rise-from-one-island-to-another/.

4. Harald Welzer, *Mentale Infrastrukturen: Wie das Wachstum indie Welt und in die Seelen kam* (Mental Infrastructures: How Growth Came into the World and into the Souls) (Berlin: Heinrich-Böll-Stiftung, 2011).

5. Riel Miller, "Changing the Conditions of Change by Learning to Use the Future Differently," *World Social Science Report 2013*, www.oecd-ilibrary.org/social-issues-migration-health/world-social-science-report-2013/changing-the-conditions-of-change-by-learning-to-use-the-future-differently_9789264203419-14-en, 107–11/.

6. Sozialistische Selbsthilfe Mülheim, "*Über* uns" (About Us) (no year), www.ssm-koeln.org/ueber_uns/intro.htm/.

7. Nico Morgenroth and Alexander Repenning, "Die Grenzen des (Post) Wachstums" (The Limits of (Post-)Growth), September 8, 2014, https://fragendreisen.wordpress.com/2014/09/08/die-grenzen-des-post-wachstums/.

8. Neal Gorenflo, Michel Bauwens, and John Restakis, "Integral Revolution: An Interview to [sic] Enric Duran about CIC," March 29, 2014,Cooperativa Integral Catalana, https://cooperativa.cat/integral-revolution/.

9. Nico Morgenroth and Alexander Repenning, "Hack the Earth," October 7, 2014, https://fragendreisen.wordpress.com/2014/10/07/hack-the-earth/.

10. Nico Morgenroth and Alexander Repenning, "Wir sind nicht die Lösung—Longo Maï" (We Are Not the Solution—Longo Maï), November 6, 2014, https://fragendreisen.wordpress.com/2014/11/06/wir-sind-nicht-die-losung-longo-mai/.

11. Grandhotel Cosmopolis, "Konzep" (Concept) (no year), https://grandhotel-cosmopolis.org/de/konzept/.

12. Quoted in Richard Saage, *Politische Utopien der Neuzeit* (Political Utopias of Modernity) (Darmstadt: Wissenschaftliche Buchgesellschaft, 1991), 1.

13. Dana Giesecke, Saskia Hebert, and Harald Welzer, *FUTURZWEI Zukunftsalmanach 2017/18. Geschichten vom guten Umgang mit der Welt. Themenschwerpunkt Stadt* (Future Perfect Future Almanac 2017/2018: Stories about Good Ways to Treat the World. Focus City) (Frankfurt am Main: Fischer Taschenbuch, 2016).

14. Transition Network, "A Movement of Communities Coming Together to Reimagine and Rebuild Our World" (2016), https://transitionnetwork.org/.

15. Goethe-Institut, "Future Perfect: Geschichten für morgen—schon heute, von überall" (Future Perfect: Stories for Tomorrow—Already Today, from Everywhere) (2019), www.goethe.de/ins/cz/prj/fup/deindex.htm/.

16. Isabelle Fremeaux and John Jordan, *Pfade durch Utopia* (Paths Through Utopia) (Hamburg: Edition Nautilus, 2012).

17. Erik Olin Wright, *Envisioning Real Utopias* (London: Verso, 2010).

18. Anwar Fazal and Lakshmi Menon, *The Right Livelihood Way: A Sourcebook for Changemakers* (Bonn: Right Livelihood College, 2016), https://rlc-blog.org/wp-content/uploads/2016/09/Prof.-Fazal-The-Right-Livelihood-Way-A-Sourcebook-for-Changemakers-1.pdf.

12 GET ORGANIZED!

1. Even their best-ever election result in 2021 remained far behind expectations—although two-thirds of Germans claim to be worried about the climate crisis, the majority did not vote for the Greens or any parties that prioritized climate in their program. www.umweltbundesamt.de/daten/private-haushalte-konsum/umweltbewusstsein-umweltverhalten#stellenwert-des-umwelt-und-klimaschutzes.

2. Annegret Kramp-Karrenbauer was successor of Angela Merkel as chair of the CDU (December 2018–January 2021), and was secretary of defense (July 2019–December 2021).

3. Translator's note: *Wende* (transformation) was the German term for the period after the fall of the Berlin wall.

4. www.agora-energiewende.de/en/.

5. www.wbgu.de/en/.

6. Erica Chenoweth and Maria J. Stephan, *Why Civil Resistance Works: The Strategic Logic of Nonviolent Conflict* (New York: Columbia University Press, 2011).

7. Hannah Arendt, *On Violence* (New York: Harcourt, 1970), 43.

8. DeCOALonize, "Lamu Residents Celebrate the Cancellation of the Lamu Coal Power Plant License," June 29, 2019, www.decoalonize.org/lamu-residents-celebrate-the-cancellation-of-the-lamu-coal-power-plant-license/.

9. Joanna Macy, *World as Lover, World as Self: Courage for Global Justice and Ecological Renewal* (Berkeley, CA: Parallax Press, 2007).

10. Gene Sharp, *From Dictatorship to Democracy: A Conceptual Framework for the Liberation* (New York: The New Press, 2002). The list doesn't stop there. The organization Nonviolence International has extended Gene Sharp's list and collected more than three hundred tactics in a database, which is accessible for free here: www.tactics.nonviolenceinternational.net/.

ABOUT THE AUTHORS AND TRANSLATOR

LUISA NEUBAUER, born in Hamburg in 1996, is among the most prominent representatives of the German climate activist movement. In 2018 she met the Swedish student Greta Thunberg at the UN Climate Change Conference and then co-founded Fridays For Future with other activists in Germany. Since then, Luisa has mobilized hundreds of thousands of people for numerous demonstrations, met various heads of state and government, and participated in the world climate conference in Madrid and Glasgow and the world economic forum in Davos. Luisa completed a bachelor's degree in geography in 2019 and is currently completing a master's in resource analysis and management at the Georg-August University in Göttingen. In July 2021 she published the book *Noch haben wir die Wahl* [We Still Have a Choice] together with journalist Bernd Ulrich of the German weekly *Die Zeit*. Her most recent book is *Gegen die Ohnmacht: Meine Großmutter, die Politik und ich* [Against Powerlessness: My Grandmother, Politics and Me], co-authored with her grandmother Dagmar Reemtsma. She is the host of the Spotify Original podcast "1.5 Grad" (1.5 degrees), and she lives in Berlin.

ALEXANDER REPENNING, born in Hamburg in 1989, is a comprehensivist, facilitator, and writer engaged for climate justice since 2015. With a bachelor's degree in social sciences from the Humboldt-University of Berlin and a master's in economics at the Cusanus Hochschule für Gesellschaftsgestaltung (Cusanus University for Shaping Society), he has been active in pushing political participation and global learning and has written about the climate crisis, postcolonial perspectives on volunteering for development, concrete utopias, and the student movement in Chile. He has published book chapters, articles, blog posts and other writings for *attac* and the blog *Postwachstum* [Degrowth]. He is currently working as education manager at Right Livelihood, the so-called Alternative Nobel Prize, connecting activism and academia and creating learning formats for system change. He lives in Annecy, France.

SABINE VON MERING, PhD (translator), is professor of German and Women's, Gender, and Sexuality Studies, and also director of the Center for German and European Studies at Brandeis University. She is a core member of the Environmental Studies Program at Brandeis and a long-time climate activist with 350Mass and NoCoalNoGas.